Theory and Applications
of Natural Language Processing

Series Editors:
Graeme Hirst (Textbooks)
Eduard Hovy (Edited volumes)
Mark Johnson (Monographs)

Aims and Scope

The field of Natural Language Processing (NLP) has expanded explosively over the past decade: growing bodies of available data, novel fields of applications, emerging areas and new connections to neighboring fields have all led to increasing output and to diversification of research.

"Theory and Applications of Natural Language Processing" is a series of volumes dedicated to selected topics in NLP and Language Technology. It focuses on the most recent advances in all areas of the computational modeling and processing of speech and text across languages and domains. Due to the rapid pace of development, the diversity of approaches and application scenarios are scattered in an ever-growing mass of conference proceedings, making entry into the field difficult for both students and potential users. Volumes in the series facilitate this first step and can be used as a teaching aid, advanced-level information resource or a point of reference.

The series encourages the submission of research monographs, contributed volumes and surveys, lecture notes and textbooks covering research frontiers on all relevant topics, offering a platform for the rapid publication of cutting-edge research as well as for comprehensive monographs that cover the full range of research on specific problem areas.

The topics include applications of NLP techniques to gain insights into the use and functioning of language, as well as the use of language technology in applications that enable communication, knowledge management and discovery such as natural language generation, information retrieval, question-answering, machine translation, localization and related fields.

The books are available in printed and electronic (e-book) form:

* Downloadable on your PC, e-reader or iPad
* Enhanced by Electronic Supplementary Material, such as algorithms, demonstrations, software, images and videos
* Available online within an extensive network of academic and corporate R&D libraries worldwide
* Never out of print thanks to innovative print-on-demand services
* Competitively priced print editions for eBook customers thanks to MyCopy service http://www.springer.com/librarians/e-content/mycopy

For other titles published in this series, go to
www.springer.com/series/8899

Antal van den Bosch • Gosse Bouma
Editors

Interactive Multi-modal Question-Answering

Springer

Editors
Antal van den Bosch
Tilburg center for Cognition and Communication
Tilburg University
School of Humanities
5000 LE Tilburg
The Netherlands
Antal.vdnBosch@uvt.nl

Gosse Bouma
Information Science
University of Groningen
9700 AS Groningen
The Netherlands
g.bouma@rug.nl

ISSN 2192-032X
ISBN 978-3-642-17524-4 e-ISBN 978-3-642-17525-1
DOI 10.1007/978-3-642-17525-1
Springer Heidelberg Dordrecht London New York

Library of Congress Control Number: 2011928374

Copy-editing: CumLingua Language & Communication/www.cumlingua.com

Cover design: deblik, Berlin

Printed on acid-free paper

Springer is part of Springer Science+Business Media (www.springer.com)

Preface

To start off with a question, what kind of book is this? This book is, in many aspects, a collaborative effort. Its contents evolve from the idea of combining the latest research in fields such as natural language processing, dialogue systems, human-machine interaction, and information extraction, around a single common theme. What is more, the book documents a successful collaborative effort to do just this: the combination of all these fields in one single research area, working towards a common goal—an interactive, multimodal spoken question-answering dialogue system, with which a human user could discuss a particular domain. This concerted effort was known as the "Interactive multimodal Information eXtraction" (IMIX) Programme, a Dutch national research programme funded by the Netherlands Organisation for Scientific Research (NWO), that ran between 2004 and 2009.

During the peak years of IMIX, the researchers in this project together formed a major part of the entire language and speech technology community in the Netherlands. The project also continued to have an influence in the work that followed, for example in the post-doc research tracks of the people who wrote their PhD thesis as part of IMIX. The majority of researchers involved in IMIX can be traced back as authors of the chapters in this book and, although the selection of chapters remains incomplete with respect to all the work performed in IMIX, we are very grateful to the many former IMIX colleagues who were willing to write retrospective overviews of their work and performance in IMIX.

The IMIX project was surrounded by a supportive network of individuals, each of whom deserves a special mention: Alice Dijkstra, Christine Erb, and Brigit van der Pas from NWO were instrumental throughout the project in organising regular IMIX meetings and, more generally, in creating an atmosphere in which everyone was encouraged to maximise performance and collaborate with as few constraints and restrictions as possible. Alice Dijkstra, who herself has a background in computational linguistics, had created a sound foundation by inviting a team of academic strategists, headed by Lou Boves, to write up the most challenging mix of ideas that, within reason, could be expected to be illustrated by the Dutch language and speech technology community, on a limited budget and within the time span of a PhD project. After an intensive assessment proposal period, the IMIX programme committee settled on the project proposals. The end results of most of these projects are covered in this book. The programme committeee appointed a project coordinator, a post occupied from the start by Els den Os, to oversee the creation of an actual running, talking, question-answering piece of demonstrator software. This book also highlights this complicated endeavour, which to some extent resembled putting together a working car from a washing machine, two lawnmowers, and a microwave oven (all ultra modern).

The project was fortunate to have the guidance of an international scientific advisory board: Eduard Hovy, Jon Oberlander, Norbert Reithinger, and Steve Young. From start to finish, their advice gave the project all the right nudges, in the right directions. Their final report now appears as the epilogue of this book. On behalf of the whole IMIX team, we would like to express our gratitude to our international advisers and to everyone who helped make this challenging project possible. For assisting us in the final editorial work, we thank Olga Chiarcos at Springer, and the team of copyeditors from CumLingua.

This book is the final publication from the IMIX project. Its aims to go beyond IMIX; it is intended as a unique look at a mix of state-of-the-art methods, rarely found together in one book (despite the fact that we are all involved in language and speech technology—such is the level of specialisation of most of that we do). But in addition to all the developments and future work born from the ideas orig in IMIX, it gives us tremendous pleasure to have had the privilege to edit this book and provide a final answer to the first question: this book is IMIX.

Tilburg and Groningen Antal van den Bosch and Gosse Bouma

Contents

Part II Interaction Management

Part III Fusing Text, Speech, and Images

Part IV Text Analysis for Question Answering

Part I
Introduction to the IMIX Programme

Introduction

Antal van den Bosch and Gosse Bouma

Abstract This book offers a broad view on the interrelationships between human-machine interaction, question answering, and spoken dialogue systems. This chapter introduces the interrelating area, Interactive Multimodal Question Answering. It also describes how this book came about: it is a reflection on a project, IMIX, that combined all of the above research fields, and offers a uniquely integrated view on how to combine all these widely differing technologies. An overview is provided of the other contributions to this book.

1 Interactive Multimodal Question Answering

Information extraction in general, and question answering (QA) in particular, is a research area that combines results from natural language processing and information retrieval to develop applications capable of advanced search over large text collections, and providing direct, detailed, and specific answers to user queries. It carries the potential to supplement or even to replace current web search engines, which concentrate on finding relevant web pages, with systems that concentrate on finding actual answers (i.e. relevant text snippets from a page) and on fusion of multiple search results in a single answer. The growing interest in this field is reflected by the fact that a number of conferences have organised workshops and special evaluation networks dedicated to this topic (the *Text Retrieval and Evaluation Conference* (TREC) and *Cross Language Evaluation Forum* (CLEF in particular), and that some of the technology has already been integrated in major search engines.

Antal van den Bosch
Tilburg University, Tilburg, The Netherlands, e-mail: Antal.vdnBosch@uvt.nl

Gosse Bouma
University of Groningen, Groningen, The Netherlands, e-mail: g.bouma@rug.nl

Yet, more was needed. The most recent activity concentrates solely on text-based retrieval, and has paid little or no attention to situations where both images and text may be required to satisfy an information need. A substantial portion of the content on the web consists of pictures and video. Web search and QA systems should be able to search all types of content simultaneously, and be able to present results to users that may consist of text, pictures, video, etc.

The emphasis on answers, answer fusion, and multimodal search results increases the need for more interactivity in the search process itself, and for input formats that not only support typing but also speech recognition and pen input. Interactive search means that a user is able to enter into a dialogue with the system to reformulate or expand queries, to zoom in or expand on search results, and to correct misunderstandings between the system and the user. Pen input allows users to ask questions about specific parts of the picture, while speech input is important to compensate for the fact that user queries about full questions and interactive dialogues require a substantial amount of input from the user. Obviously, speech input may also be a bonus on mobile devices.

2 The IMIX Project

The *Interactive Multimodal Information Extraction* (IMIX) project was initiated by the Netherlands Organisation for Scientific Research (NWO) to foster interdisciplinary cooperation between various research groups active in the fields of speech recognition and synthesis, dialogue management, multimodal user interfaces, and NLP in general. The programme consisted of seven research projects. All projects worked on the same application domain of medical QA, and contributed to a single common demonstrator. An up-to-date web portal containing full information on the project, as well as a complete list of publications, is maintained by NWO at `www.nwo.nl/imix`.

The development of novel, multimodal and interactive, search, information extraction, and QA systems requires input from several disciplines. Even though many of these disciplines are closely connected, and at times use results from neighbouring fields, communication and true cooperation between groups remains a challenge. The IMIX project was no different: it was an interesting and challenging experiment. While primarily aimed at funding high quality research by individual groups, it also encouraged these groups to cooperate on advanced multimodal search applications. This book presents some of the most important lessons learned during the project.

3 Contributions

The contributions in this book address issues relevant to the development of an interactive, multimodal, medical QA system. Furthermore, some contributions also touch on more general issues in NLP, such as the development of multidimensional dialogue systems, the acquisition of taxonomic knowledge from text, and part-of-speech(POS) tagging. Together, they give an overview of the most important results within the IMIX project.

The rest of this book is organised as follows:

Introducing the IMIX project

This chapter is followed by a second introductory chapter, *The IMIX demonstrator: an information search assistant for the medical domain* by Dennis Hofs, Boris van Schooten, and Rieks op den Akker. It describes the implementation of the demonstrator system. The IMIX Demonstrator developed into a complex application that had to integrate several modalities (speech recognition and synthesis, text and pen input, display of text and visual search results, discourse management for text, speech, and an avatar), yet in the end was made to work as a spoken interactive question-answering dialogue system, thanks to well-chosen design strategies and the choice of a particular real-world domain: the medical domain. The chapters in the remainder of the book all address research that contributed more or less directly to the Demonstrator. In addition, Hämäläinen et al (2005) and Han et al (2005) report on work on speech recognition within IMIX (carried out in the NORISC project).

Interaction Management

Two complementary chapters are dedicated to dialogue management. Together, they provide an answer to the questions:- What does it take to practically build a dialogue system for an interactive and multimodal, QA system? What theoretical issues need to be addressed to ensure that users can actually interact with the system in a natural fashion?

Corpus-based Development of a Dialogue Management System for Multimodal Question Answering by Boris van Schooten and Rieks op den Akker presents the result of efforts to create corpora of question-answering dialogues for multimodal information retrieval. While corpora for purely textual non-interactive QA have proved to be extremely important for the evaluation and development of corresponding QA systems, far less data is available for multimodal, interactive, QA. This chapter describes three corpora created within the project, and how these were used in the design of the dialogue manager. It describes clearly how a corpus-based approach can be used to arrive at a state-of-the-art dialogue manager.

Multidimensional Dialogue Management by Simon Keizer, Harry Bunt, and Volha Petukhova addresses the fact that user and system utterances in a dialogue manager often combine different dialogue acts. Users may ask a (follow-up) question about the domain, while at the same time signalling that they accepted some previous system utterance. Systems may ask for clarification while at the same time giving feedback on the interpretation of the previous user utterance by the system. The authors present a dialogue manager that is capable of processing utterances among different dimensions in order to support natural dialogue.

Fusing text, speech, and images

Two chapters concentrate on the question of how to present the user with information that can be a combination of text, speech and images, extracted by means of a QA, or an information extraction component.

Experiments in multimodal information presentation by Charlotte van Hooijdonk, Wauter Bosma, Emiel Krahmer, Alfons Maes, and Mariët Theune describe a search for the optimal interplay between text and images in presenting medical information.

Text-to-text generation for Question Answering by Wauter Bosma, Erwin Marsi, Emiel Krahmer, and Mariët Theune presents sentence fusion and generation methods for combining multiple text snippets retrieved by a QA system into a single text, to be presented as spoken answers to the user.

Text analysis for Question Answering

There are five chapters dedicated to various aspects of QA and information extraction for the medical domain. They are all concerned with enriching the baseline QA system by extending the basic feature set, the words in the queries and documents, with syntactic, semantic, and domain-specific textual features. An interesting tension can be observed between (i) the supervised machine-learning approach, that relies on annotated data and thus produces domain-specific analysis modules, (ii) approaches that infer taxonomic lexical semantic relations in an unsupervised manner, from either domain-specific or domain-independent data, and (iii) the approach that relies on domain-independent, generic, but language-specific NLP. The conclusion is that each approach has its benefits, but drawbacks are also clearly outlined and discussed.

Automatic Extraction of Medical Term Variants from Mutilingual Parallel Translations by Lonneke van der Plas, Jörg Tiedemann, and Ismail Fahmi addresses the fact that there often exists a 'lexical gap' between the terminology used by users and used in medical documents. The authors present a system based on alignment of words in multilingual parallel corpora for detecting term variation and synonymy.

Relation Extraction for Open and Closed Domain Question Answering by Gosse Bouma, Ismail Fahmi, and Jori Mur discusses methods for information extraction based on syntactic dependency relations and semantic concept labels. The methods are tested in an open QA system and in a closed-domain (medical) QA system, implementing one of the IMIX modules.

Constraint-Satisfaction Inference for Entity Recognition by Sander Canisius, Antal van den Bosch, and Walter Daelemans describes techniques for sequence learning, and applies these, and others, to the problem of detecting and labelling domain-specific named entities in medical text.

Extraction of Hypernymy Information from Text by Erik Tjong Kim Sang, Katja Hofmann, and Maarten de Rijke investigates and compares several techniques for finding hypernym concept pairs (such as *influenza — disease*).

Towards a Discourse-driven Taxonomic Inference Model by Piroska Lendvai presents a system that combines the acquisition of taxonomic relations (such as hypernyms) using paragraph structure, with learning information extraction patterns based on medical concept labels.

Epilogue

The final chapter, *IMIX: Good Questions, Promising Answers* by Eduard Hovy, Jon Oberlander, and Norbert Reithinger, provides a critical evaluation of the results of the various subprojects within IMIX, and positions the results in a wider, international, context. Furthermore, it contains suggestions for future, follow-up research.

4 Further Reading

The IMIX project ended effectively in 2008. After completion of the project, several contributors have continued their research on topics closely related to the work done within IMIX. We end this chapter with a selected overview of follow-up work arising from the IMIX project.

The work on dialogue management described, in the chapter by Van Schooten en Op den Akker resulted in a study on handling follow-up questions in dialogue managers (Van Schooten et al, 2009). The work by Keizer, Bunt, and Petukhova was followed up by work on the annotation of multifunctionality in dialogue (Petukhova and Bunt, 2009; Bunt, 2011).

The work on multimodal information presentation and generation was followed up by work on query-based sentence fusion (Krahmer et al, 2008), and on the influence of visuals in text on the learning and execution of repetitive strain injury exercises (Van Hooijdonk and Krahmer, 2008).

In the area of text analysis and QA, Ittoo and Bouma (2010) describes an approach to relation extraction that builds on the work outlined in the chapter by Bouma, Fahmi, and Mur. It uses large automatically parsed corpora to learn dependency patterns for extracting instances of a given relation. The lexical acquisition work, described in the chapter by van der Plas, Fahmi, and Tiedemann, is continued in Van der Plas et al (2010), which investigates the acquisition of French synonyms from parallel corpora. The approach to acquistion of hypernyms, advocated in the chapter by Tjong Kim Sang, Hofmann, and De Rijke, is explored for other lexical relations (i.e. synonyms and antonyms) in Lobanova et al (2009) and Lobanova et al (2010).

The work on structured output learning by Canisius, Daelemans, and Van den Bosch was further applied to dependency parsing (Canisius and Tjong Kim Sang, 2007) and machine translation (Canisius and Van den Bosch, 2009). Further work on extracting taxonomic domain knowledge from semi-structured documents such as encyclopedias was extended to expedition field books (Van den Bosch et al, 2009).

References

Van den Bosch A, Lendvai P, Van Erp M, Hunt S, Van der Meij M, and Dekker R (2009) Weaving a new fabric of natural history. Interdisciplinary Science Review 34(2–3):206–223.

Bunt, H (2010) Multifunctionality in dialogue. Computer Speech & Language 25(2):222–245.

Canisius S, Tjong Kim Sang E (2007) A constraint satisfaction approach to dependency parsing. In: Proc. of the CoNLL Shared Task Session of EMNLP-CoNLL 2007, Prague, Czech Republic, pp 1124–1128

Canisius S, Van den Bosch A (2009) A constraint satisfaction approach to machine translation. In H. Somers and L. Màrquez (Eds.), Proceedings of the 13th Annual Conference of the European Association for Machine Translation, Barcelona, Spain, pp 182–189

Hämäläinen A, Boves L, De Veth J (2005) Syllable-length acoustic units in large-vocabulary continuous speech recognition. In: Proceedings of SPECOM, pp 499–502

Han Y, De Veth J, Boves L (2005) Speech Trajectory Clustering for Improved Speech Recognition. In: Ninth European Conference on Speech Communication and Technology

Van Hooijdonk C, Krahmer E (2008) Information modalities for procedural instructions: the influence of text, static and dynamic visuals on learning and executing RSI exercises. IEEE Transactions on Professional Communication 51(1):50–62

Ittoo A, Bouma G (2010) On learning subtypes of the part-whole relation: do not mix your seeds. In: Proceedings of the 48th Annual Meeting of the Association for Computational Linguistics, Association for Computational Linguistics, pp 1328–1336

Krahmer E, Marsi E, Van Pelt P (2008) Query-based Sentence Fusion is Better Defined and Leads to More Preferred Results than Generic Sentence Fusion. In: Proceedings of ACL-08: HLT, Short Papers, Columbus, Ohio, pp 193–196

Lobanova A, Spenader J, Van de Cruys T, Van der Kleij T, Tjong Kim Sang E (2009) Automatic Relation Extraction-Can Synonym Extraction Benefit from Antonym Knowledge? In: Proceedings of WordNets and other Lexical Semantic Resources - between Lexical Semantics, Lexicography, Terminology and Formal Ontologies, Odense, Denmark, pp 17–20

Lobanova A, Bouma G, Tjong Kim Sang E (2010) Using a Treebank for Finding Opposites. In: Dickinson M, Müürisep K, Passarotti M (eds) Proceedings of the ninth Workshop on Treebanks and Linguistic Theory, Tartu, Estonia, pp 139–150

Petukhova V, Bunt H (2009) The independence of dimensions in multidimensional dialogue act annotation. In: Proceedings of Human Language Technologies: The 2009 Annual Conference of the North American Chapter of the Association for Computational Linguistics, Companion Volume: Short Papers, pp 197–200

Van der Plas L, Tiedemann J, Manguin J (2010) Automatic acquisition of synonyms for French using parallel corpora. In: Proceedings of the 4th International Workshop on Distributed Agent-Based Retrieval Tools

Van Schooten B, Op den Akker R, Rosset S, Galibert O, Max A, Illouz G (2009) Follow-up question handling in the IMIX and Ritel systems: A comparative study. Natural Language Engineering 15(01):97–118

The IMIX Demonstrator: An Information Search Assistant for the Medical Domain

Dennis Hofs, Boris van Schooten and Rieks op den Akker

Abstract In the course of the IMIX project a system was developed to demonstrate how the research performed in the various subprojects could contribute to the development of practical multimodal question answering dialog systems. This chapter describes the IMIX Demonstrator, an information search assistant for the medical domain. The functionalities and the architecture of the system are described, as well as its role in the IMIX project.

1 Introduction

The IMIX research programme covered four research areas:

(i) Automatic speech recognition (ASR) within the context of multimodal interaction;
(ii) Dialogue management and reasoning;
(iii) Information presentation in multimodal systems in natural language;
(iv) Information extraction.

An important aim of the IMIX Programme was to bring research from these different fields together. The building of a common demonstrator was seen as a means to this end. This chapter describes the common demonstrator. The challenge was to find an appropriate architecture which would allow the development of loosely coupled modules in such a way, that changes to one module during

Dennis Hofs
Roessingh Research and Development, Enschede, The Netherlands, e-mail: d.hofs@rrd.nl

Boris van Schooten
University of Twente, Enschede, The Netherlands, e-mail: schooten@ewi.utwente.nl

Rieks op den Akker
University of Twente, Enschede, The Netherlands, e-mail: infrieks@ewi.utwente.nl

the course of the project would not influence the set up of the whole system. Section 3 presents the architecture and describes the different modules contributed by the various subprojects. The next section presents the functionality of the IMIX Demonstrator and the interaction with the user. The final section 4 discusses the role the Demonstrator played in the IMIX project. The technical coordination of the Demonstrator was done by the first author of this chapter. He was responsible for the integration of the various modules. The kernel of the Demonstrator is the dialogue manager module. It is dependent on every module, so architecture and functional specification was managed by the Vidiam subproject (see the chapter by Van Schooten and Op den Akker, *Vidiam: Corpus-based Development of a Dialogue Manager for Multimodal Question Answering*, this volume). The Demonstrator was developed in three phases in the course of three years. This book describes the final version of the Demonstrator[1].

2 A Medical Information Search Assistant

IMIX is an information assistant, a system that helps users find the information they need in the medical domain (op den Akker et al, 2005). The IMIX Programme extends iterative QA in the direction of multimodal QA. IMIX allows users to use speech input as well as text and gestures. The answer is presented in the form of text, speech and pictures. Users are allowed to refer to the pictures in the answer presentation when asking follow-up questions. They can use pointing gestures, encircle parts of pictures or words, or underline them. IMIX is able to offer users a list of options, from which they can make a selection, including using pointing.

Particular choices have been made regarding the nature and difficulty level of the questions IMIX is supposed to handle. An attempt has been made to provide coverage that is sufficient to answer the information needs of real, non-expert users. The questions this project covers have been called “encyclopedic questions”. These are general medical questions that do not require expert medical knowledge (this excludes diagnostic questions). Nevertheless, correctness of answers is less well-defined and harder to measure than with factoid questions. The system typically answers with a piece of text about 1-3 sentences long. The length and level of detail of a correct answer depends on the information needs of the user, which cannot be precisely known beforehand. Enabling the user to give feedback on their information need is one of the purposes of the dialogue system. Although IMIX focuses on a closed medical domain, many of the techniques used are also applicable to open domain QA. This is especially true of the dialogue management module. It uses minimum domain knowledge, which makes it easy to generalise to other domains.

The user can enter a question to the system in Dutch, either by typing or speaking. Both input modes can be combined with pen input (pointing or drawing). In a follow-up question, the user can use the mouse or a pen to encircle parts of a

[1] A video presentation of the IMIX Demonstrator can be found on `www.youtube.com/watch?v=swA8_Y56aak`

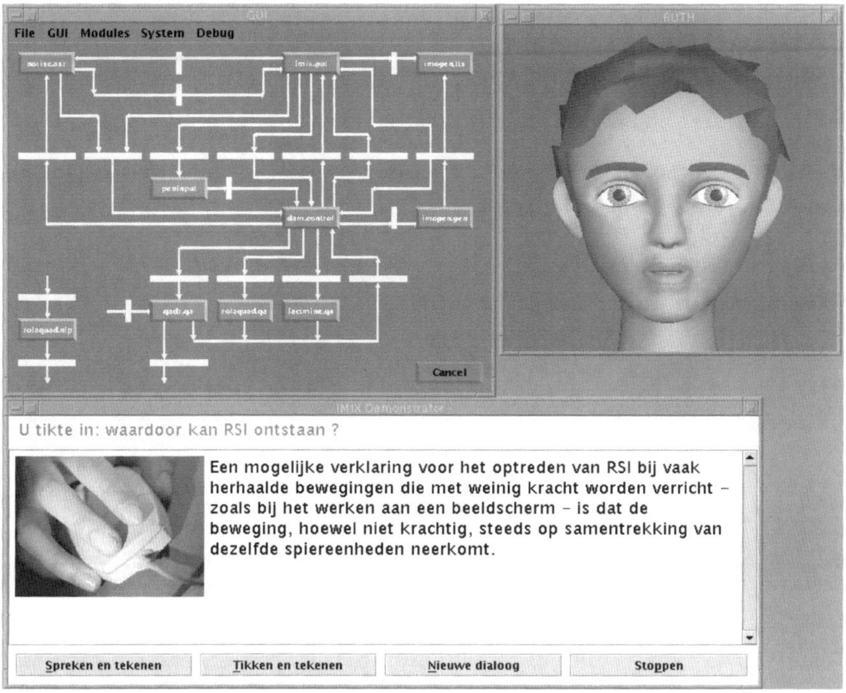

Fig. 1 A snapshot showing parts of the user interface of the IMIX Demonstrator. The left-upper part shows the control GUI that comes with the MULTIPLATFORM. Boxes are system modules; they light up when activated. The right corner shows RUTH, the talking head. The buttons allow the user to select the input modality, speech or text, and to signal the start of a new dialogue.

picture when he wants to ask a question about it (see Figure 2). Follow-up questions may contain anaphoric references ("How do I recover from it?") that need to be rewritten to form a self-contained question ("How do I recover from RSI?") that can be handled by a QA engine. To a limited extent IMIX can handle utterances that are out of domain, in particular those that express how the user assesses the answer IMIX gave to a previous question. For the different types of utterances that IMIX can handle, consult the chapter by Van Schooten and Op den Akker, *Vidiam: Corpus-based Development of a Dialogue Manager for Multimodal Question Answering*, this volume.

A production experiment was carried out in the IMOGEN project to determine which modalities people choose to answer different types of questions. In this experiment participants had to create (multimodal) presentations of answers to general medical questions. The collected answer presentations were coded to types of manipulations (typographic, spatial, graphical), presence of visual media (i.e., photos, graphics, and animations), functions and the position of these visual media. The results of a first analysis indicated that participants presented the information in a multimodal way (van Hooijdonk et al, 2007).

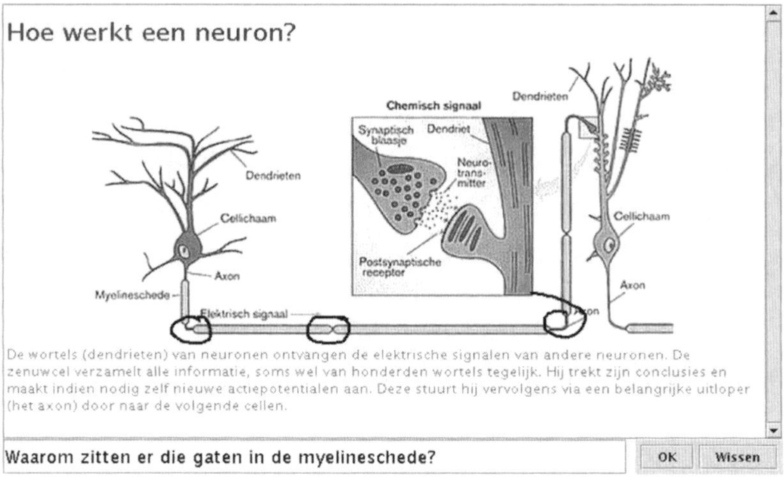

Fig. 2 An example of a multimodal follow-up question. At the top is the original question (translated: "How does a neuron work?") and in the middle is the answer. In the box at the bottom is the user's follow-up question (translated: "Why are there these holes in the myelin sheath?"). The circles in black are the mouse strokes made by the user.

The information assistant is personified in the form of a Dutch-speaking version of the Rutgers University Talking Head (RUTH[2]). In the Demonstrator shown in the movie on YouTube the Festival TTS system is used. Figure 1 shows a screen capture of the user interface of IMIX. Presentation of the information assistant by means of a talking head or an embodied conversational agent was not the focus of one of the IMIX research projects. It is well known that adding a character on the interface, it does not matter how primitive the figure is, has an impact on the way the user expresses their information requirement. If there is a face, users are more inclined to formulate full sentences (full questions) instead of only keywords, as they often do when the system lacks a human-like interface.

For the motivation behind the choice of the medical domain see Boves and den Os (2005), which also presents a short overview of an earlier version of the IMIX Demonstrator.

3 Architecture of the Final Version

This section describes IMIX's architecture. The interaction with the user that it supports is best explained by the graphical user interface. Figure 3 gives a schematic impression of the various screens of the user interface of IMIX. The GUI starts with

[2] www.cs.rutgers.edu/~village/ruth/

a welcome screen (1) with a Start button that takes the user to screen 2. At the start of a dialogue that screen shows some initial information and instructions. The user can choose to speak or type a question, or to stop IMIX.

The first time speech input is chosen, the ASR module needs to be initialised (screen 3). Then, the screen shows that the user can speak. If no speech is detected before a time-out, the GUI will go back to screen 2. Otherwise, the recognised speech is shown in screen 6, while the DAM is processing it. For speech input, the user can enter a question in screen 5, which will also end in screen 6.

When the Dialogue and Action Manager (DAM) and other modules have finished processing the user input and an answer has been generated, the GUI goes back to screen 2, but now showing the answer rather than the start-of-dialogue text.

Now, when the user chooses to speak or type a new question, the last answer remains visible and the GUI allows the user to draw (encircle, underline) in the answer. It is also possible to only draw, without any textual input.

In screen 2, the user can start a new dialogue at any time. IMIX includes the function to let the user fill in a survey after completing a dialogue (screen 7). The dialogue act options shown in screen 2 are a function to facilitate development of the DAM, and are not visible in the final Demonstrator.

IMIX is implemented using MULTIPLATFORM, an integration platform based on a general framework for building integrated natural-language and multimodal dialogue systems. MULTIPLATFORM was developed by DFKI in the VerbMobil and the SmartKom (Wahlster, 2006) projects and was used in COMIC, a European project on multimodal dialogue systems (Catizone et al, 2003) to develop integrated system prototypes (Herzog et al, 2004). The framework supports the modularisation of the dialogue system into distinct and independent software modules to allow maximum decoupling of modules, a facility that is urgently needed in a project like the IMIX project, with largely autonomous partner projects that deliver their modules for the Demonstrator. Communication between distributed software modules can be implemented in various languages and on distributed machines. This is realised by a publish/subscribe approach. The sender publishes a notification on a named message queue, so that the message can be forwarded to a list of subscribers. This kind of distributed event notification makes the communication framework very flexible as it focuses on the data to be exchanged and it decouples data producers and data consumers. A named message queue is called a data pool. Every data pool can be linked to an individual data type specification to define admissible message contents. The platform includes a control GUI that shows the activated modules and their connections (see the left-upper part of Figure 1.)

Figure 4 shows a detailed architecture of the last version of IMIX. Rounded rectangles denote data pools of MULTIPLATFORM, the grey boxes are modules. Shared data stores, ontologies, encyclopedia and image databases are shown in the bottom right of the picture. The next section introduces the functional modules of the architecture.

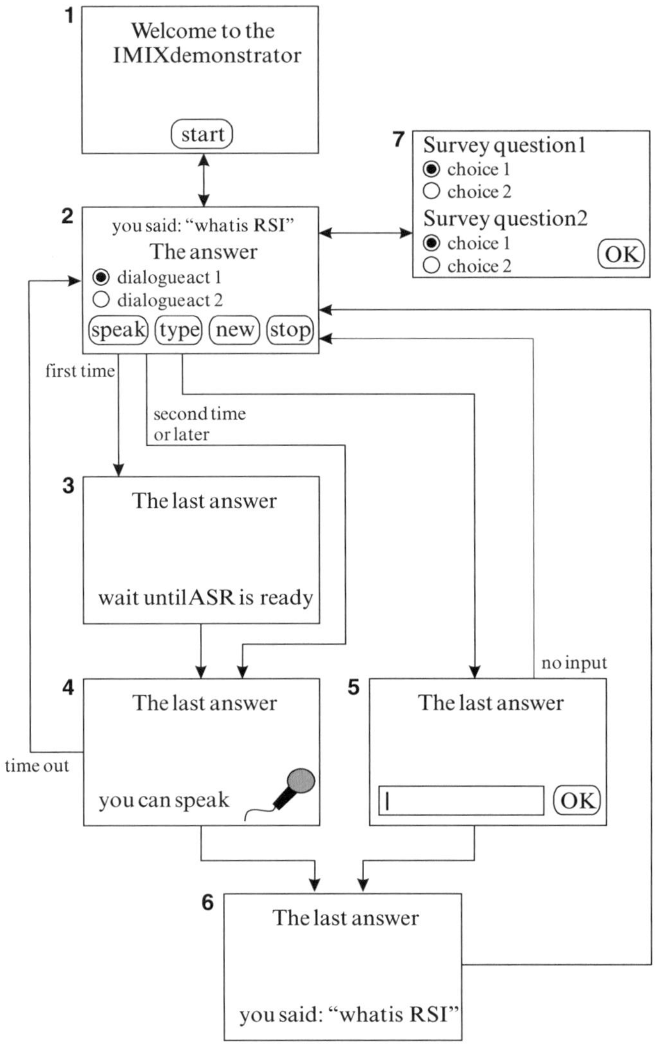

Fig. 3 Schematic overview of the various screens of the user interface.

3.1 The Modules

The main module is the Dialogue and Action Manager, DAM.Control. It has connections with all other modules, the GUI, ASR, the four QA modules, TTS and GEN.

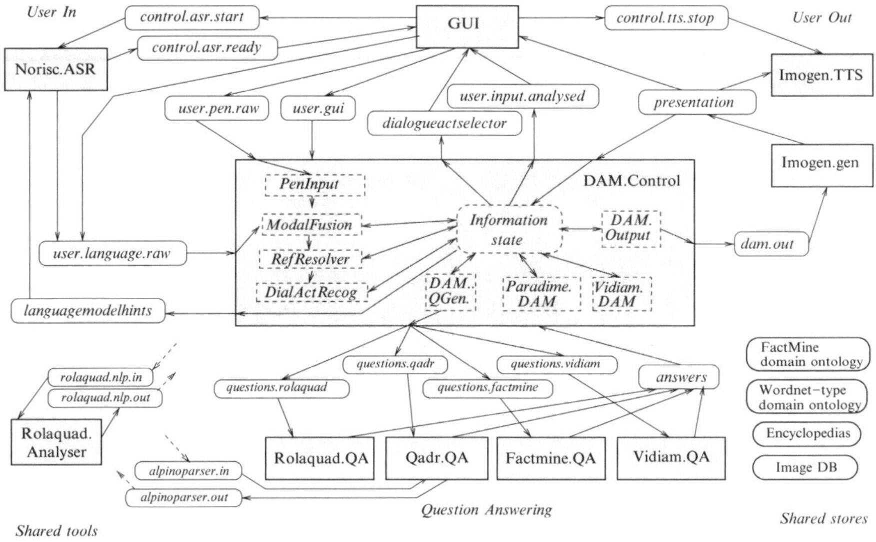

Fig. 4 A detailed architecture of the third and last version of the IMIX Demonstrator.

DAM.Control. DAM.Control consist of several submodules that process the multimodal pen/text/speech input data in the order shown in the architecture: PenInput, ModalFusion, RefResolver and DialActRecog. They communicate via the Information State, which holds the information about the dialogue acts. Feature structure unification is used for completing elliptic follow-up questions and for solving anaphoric references by the RefResolver.

PenInput matches pen actions with annotated visual areas. It can handle visual areas and aggregates of areas, arrows, and underlining. The module was tested with the multimodal follow-up question corpus and annotated images with good result (88% of identifiable visual elements found). The RefResolver resolves visual referents using the input from PenInput, combined with properties of the visual elements on the screen (in particular, colour, shape, name, and function).

DialActRecog tries to classify the type of speech act that the user performs (see the Vidiam chapter for the various types of acts that are distinguished.)

After the analytical steps the DAM module decides what system action will be executed next. Paradime DAM and Vidiam DAM are interchangeable for this function. The chosen system action is either a question to be sent to the QA engines or another dialogue act to be sent to the user directly. The output is sent to DAMOutput.

Norisc.ASR. ASR is provided by the Norisc.ASR module, based on HTK (Young et al, 2006). It is only used when speech input is switched on. The ASR acoustic models and language models are trained for the specific RSI subdomain to get maximum performance.

QA modules. There are four QA modules. They all share the same interface with the DAM: question in and retrieved results out. Rolaquad.QA is based on machine learning and semantic tagging of words using an ontology of medical terms. QADR is a QA system that exploits full parsing of the document and questions (Bouma et al, 2005). It uses the Alpino dependency parser. FactMine.QA is based on an unsupervised learning method. Vidiam.QA uses a simple TF.IDF based search method.

Imogen Output. The QA engines return a ranked list of text snippets, usually consisting of one sentence, that should contain an answer to the question. The DAM passes this list to the Imogen module, which generates the answer text that is presented to the user. It decides which of the potential answers to use and how to extend this 'core' answer so that the output is a comprehensive and verifiable answer to the question. TTS runs with the talking head RUTH in one module.

3.2 The DAM State Machine

The DAM state diagram (see Figure 5) describes the sequencing of messages as input and output by the DAM. This largely determines the control flow of the entire application, as all modules except the GUI simply reply to the DAM's messages. In WaitUser, the DAM waits for the GUI to pass the user utterance. Once the GUI messages are received, the system goes to ProcessInput. Here, the DAM calls the appropriate taggers and analysers to analyse the input as necessary, after which the GenerateAct state is reached. Then it calls the QA engines if necessary, and generates the output, after which it reaches OutputAct. As soon as Imogen.gen has finished generating the final presentation, the cycle repeats. There is a special state called OfflineProcessing, which is active when an "offline processing" flag is set at startup. This enables the DAM components to be tested using offline data.

3.3 A Modular Version of the Demonstrator

The version of IMIX implemented on MULTIPLATFORM runs on a Linux machine. This version did not allow remote access. To make the IMIX system accessible for multiple users a modular multi-client server configuration was developed. The Dialogue Manager module and the GUI now run on the client machine that is connected with the MUP server via TCP. Various TTS or ASR systems can be connected to the client application.

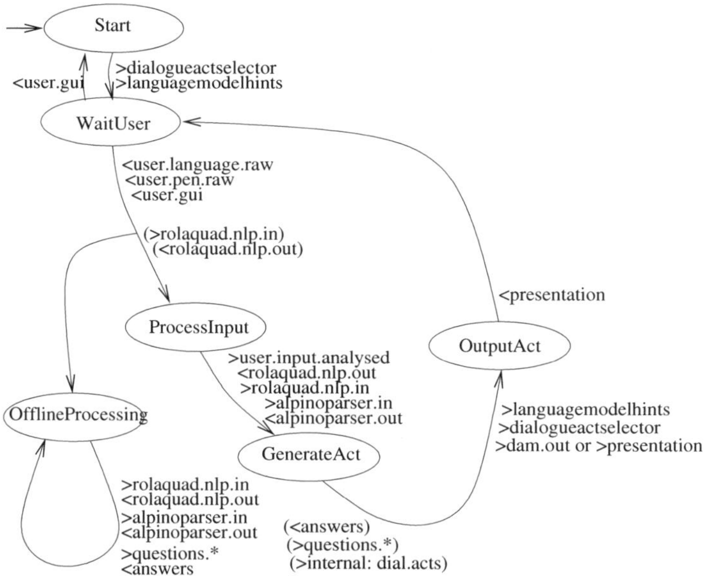

Fig. 5 The DAM state diagram. The ellipses represent the states, the arrows the transitions. Alongside the arrows are the pools that are read from or written to. The $<$ sign indicates reading, $>$ indicates writing. The order in which the operations are shown is the order in which they take place. Read operations block the DAM until the message is received.

4 Conclusion

The IMIX project developed a working system that satisfies its functional requirements for demonstrative purposes. The Demonstrator also makes clear where the missing bits are, that stand in the way of a practical multimodal information search assistant. Therefore, good and fast retrieval systems for multimodal information are the least that is needed, a topic that was not properly covered by the IMIX Programme. This requires automatic recognition of visual elements and automatic annotation of parts of illustrations, as well as research into how text and visual elements are connected in media. The annotations are also needed for the user interacting with the system through pointing. More research is needed to see how people refer to visual elements and how this is related to their speech, and the dialogue context. More research is needed in generation of speech, gestures and expressions (audible and visible) of the mental (cognitive, affective) state of the assistant to make the assistant a more believable character, and give it a more sophisticated face. And finally, ASR and dialogue researchers should work more closely together to come up with improved models of dialogue and context dependent speech recognisers. Domain knowledge is a prerequisite for a practical system, that will always be focussed on a specific domain. It will be clear that for a dialogue system with "natural" turn-taking not only effectiveness of these technologies is important; they must also be fast.

This chapter concludes with a few words about the various roles the Demonstrator played in the project.

Demo systems definitely have a function in the relationship between the research community and the broader public. For example, in engaging interested high school students in research in a particular area of research and technology. People generally like to see something demonstrable. According to the self evaluation report by the Programme Committee "the IMIX demonstrator was never meant to be an early version of a prototype of an operational application. On the contrary, the goal of the common demonstrator was to provide hard and fact evidence of the capability of the research teams funded through IMIX to collaborate on a level that would allow the integration of most of the results in a comprehensive system. The possibility of using the common demonstrator as a means for attracting attention from the general public and thereby increasing the interest in applications of language and speech technology for commercial and social purposes was only an ancillary goal." Clearly, the Demonstrator had an *internal* function: to have partners from different disciplines collaborate, not just present their work in a joint workshop, but have them deliver their software in a way that "allows integration in a comprehensive system." Certainly in the first half of the project period the regular Demonstrator meetings had a positive impact on the communication between the research groups. Towards the end of the project there were clearly other priorities, PhD students had to finish their thesis and postdocs had to apply for other jobs. The second internal role of the Demonstrator was that of a research subject itself. It is clear that for some of the subprojects in IMIX the Demonstrator was more important than for others. For the Vidiam dialogue manager project a working Demonstrator was the only way to evaluate the dialogue manager as a complete system. The architecture of the Demonstrator had a strong impact on the DAM.Control module. The QA engines and the DAM.Control modules are now loosely coupled: the DAM sends a question to the QA engine and gets the results from it. This reflects the "modular" structure of the project as a whole, with its highly autonomous subprojects. Ideally the DAM-QA interface would be much broader and they would even share the dialogue information state, to properly handle follow-up questions. The same is true of the connection between the ASR and the dialogue modules. Ideally, the speech recognition is informed by the dialogue state which induces prior word expectations.

The Demonstrator software was hardly used for research, for example to collect dialogues. This is also the case with the individual modules. Most collaboration between subprojects was not related to the Demonstrator, but to the use of supplementary resources, such as NLP tools (for example the use of the Alpino parser) and shared data, such as the collections of annotated questions and annotated pictures. All in all, the conclusion is that the Demonstrator itself did not work as a research tool. So, was the Demonstrator a good idea? It served as a collaboration tool, but there may be more efficient methods.

References

op den Akker HJA, Bunt HC, Keizer S, van Schooten BW (2005) From question answering to spoken dialogue - towards an information search assistant for interactive multimodal information extraction. In: Proc. 9th European Conference on Speech Communication and Technology (Interspeech 2005), Lisbon, Portugal, European Speech Communication Association (ESCA) / CEP Consultants, pp 2793–2796

Bouma G, Mur J, van Noord G, van der Plas L, Tiedemann J (2005) Question answering for dutch using dependency relations. In: Working notes for the CLEF 2005 workshop, Springer

Boves L, den Os E (2005) Interactivity and multimodality in the imix demonstrator. IEEE International Conference on Multimedia and Expo 0:1578–1581

Catizone R, Setzer A, Wilks Y (2003) Multimodal dialogue management in the COMIC project. In: Proceedings of the EACL 2003 Workshop on Dialogue Systems: Interaction, Adaptation, and Styles of Management, Budapest, Hungary

Herzog G, Ndiaye A, Merten S, Kirchmann H, Becker T, Poller P (2004) Large-scale software integration for spoken language and multimodal dialog systems. Natural Language Engineering 10 (3/4):283–305

van Hooijdonk C, Krahmer E, Maes F, Theune M, Bosma W (2007) Towards automatic generation of multimodal answers to medical questions: a cognitive engineering approach. In: Proceedings of the Tenth International Symposium on Social Communication

Wahlster W (ed) (2006) SmartKom: foundations of multimodal dialogue systems. Springer

Young S, Evermann G, Gales M, Hain T, Kershaw D, Liu XA, Moore G, Odell J, Ollason D, Povey D, Valtchev V, Woodland P (2006) The HTK Book. Cambridge University Engineering Department

Part II
Interaction Management

Vidiam: Corpus-based Development of a Dialogue Manager for Multimodal Question Answering

Boris van Schooten and Rieks op den Akker

Abstract This chapter describes the Vidiam project, which covered the development of a dialogue management system for multimodal question answering (QA) dialogues, as carried out in the IMIX project. The approach followed was data-driven, i.e., corpus-based. Since research in QA dialogue of multimodal information retrieval is still new, no suitable corpora were available to base a system on. This chapter reports on the collection and analysis of three QA dialogue corpora, involving textual follow-up utterances, multimodal follow-up questions, and speech dialogues. Based on the data, a dialogue act typology was created, which helps translate user utterances to practical interactive QA strategies. The chapter goes on to explain how the dialogue manager and its components: dialogue act recognition; interactive QA strategy handling; reference resolution; and multimodal fusion, were built and evaluated using off-line analysis of the corpus data.

1 Introduction

The Vidiam project (DIAlogue Management and the VIsual channel) aims at combining multimodal unstructured QA technology with natural language dialogue. It is well known that, in order to be feasible, dialogue systems can cover only a limited domain, to enable typical dialogues within the domain to be modelled explicitly. Most dialogue systems follow a "slot-filling" approach to managing the dialogue, which amounts to having the user fill in predetermined values. There are relatively few dialogue systems which can handle dialogue for an underlying QA engine. Most of these systems only have a limited concept of QA dialogue, so a comprehensive coverage of the potential of QA dialogue is still a long way off.

Boris van Schooten
University of Twente, The Netherlands, e-mail: `schooten@ewi.utwente.nl`

Rieks op den Akker
University of Twente, The Netherlands, e-mail: `infrieks@ewi.utwente.nl`

The Vidiam project concentrated mainly on handling user follow-up utterances (FU). These are any user utterances given after the system produces (or fails to produce) an answer. A major class of FU are follow-up questions (FQ). These are regular QA questions, except that they are incomplete, and can only be understood as part of previous utterances in the dialogue.

The possibilities of FU handling were explored by bringing together two perspectives: the user perspective (what do users want and how do they act?) and the technical perspective (what is currently technically possible?). The technical perspective provided in Section 2 involves a review of current QA dialogue systems, and an overview of the FU handling techniques that are known to work. The user perspective involves collecting corpora of QA dialogues and user FU. The two perspectives are brought together by classifying the utterances in the corpora according to what type of processing and response which is required by the system, based on the range of known technical possibilities. This is covered in Section 3.

In addition to corpus analysis, work was also carried out on two dialogue managers (DAMs): the Vidiam DAM, which is a module in the IMIX demonstrator, and dialogue management functionality for the Ritel system, which was developed in a 5-week collaboration with the Ritel team (Galibert et al, 2005). The Vidiam DAM is based on the FU classification, and Section 4 evaluates how well the classification and subsequent FU handling performs. The Ritel system was chosen because it complements the IMIX system in terms of features (see also Section 2). In particular, it is speech-based. Also, in Ritel it was possible to collect a corpus of full dialogues with the real system. The analysis of Ritel concentrates on speech, and is mostly described in Section 3.3. Speech was never implemented in the Vidiam DAM because of the lack of appropriate ASR, and hence, there is no evaluation of speech handling in the Vidiam DAM evaluation.

Most of the findings are desribed in previous work (van Schooten and op den Akker, 2005; van Schooten et al, 2009, 2007; van Schooten and op den Akker, 2007). This chapter summarises the analysis of the different corpora and reflects on the consequences for the design of a QA dialogue system.

1.1 QA Dialogue System Features

The manner in which FU can or should be handled depends on the features of a particular QA system. The QA system features are characterised along two dimensions: the available modalities, and the ability to answer different types of questions. It is within this framework that IMIX, Ritel and the systems from the literature will then be placed.

1.1.1 Available Modalities

Most QA systems only handle text (typed) questions, answered by text answers. A minority of systems, such as Ritel, handle speech input, which presents quite a different challenge, because of the speech processing problems of large-vocabulary ASR. Few QA systems, including the IMIX and SmartWeb (Reithinger et al, 2005) systems, handle multimodal FQ to multimodal answers.

Speech.

ASR output for QA is typically so noisy, that it can be expected that something like half of the output will be unusable. The most serious problem with the current Ritel system lies in the quality of ASR, with a word error rate of about 30%, and the error rate of keywords being around 50% (van Schooten et al, 2007). The IMIX ASR was not even tested with a sufficiently large vocabulary. Speech requires repair dialogue, which is treated as a separate subdialogue in Ritel, and is therefore independent of FQ handling. However, speech may influence user behaviour, and may make certain FQ more, or less common or desirable. For example, anaphoric FQ in particular would have added value in speech QA, as they reduce the need for repeating the keywords that are so difficult to recognise by an ASR.

Multimodality.

Multimodality is defined here as the combination of the presentation of multimodal answers (text + pictures) with multimodal FQ (text/speech + pointing or gesturing on the screen). Multimodality requires an extra interpretation and/or generation step in the QA process.

Handling of multimodal FQ is rarely found, but there are some multimodal "QA" dialogue systems which do not use unstructured QA technology, but a more knowledge-rich approach. Examples of these are found in Wang et al (2006), a Chinese storytelling system; Martin et al (2006), the Andersen storytelling system; and Hofs et al (2010), a navigation dialogue system.

It is assumed that multimodal FQ can be handled in basically the same way as unimodal ones, except that there are specific classes of FQ that only exist in the multimodal case. In van Schooten and op den Akker (2007), almost all multimodal FQ in the corpus were found to be primarily linguistic. The pointing gestures were used only to disambiguate the references made in the linguistic component of the utterances. This is mostly consistent with findings in the Andersen system, which found only 19% gesture-only user turns. These could be interpreted as simply "What is ...?" queries.

1.1.2 Ability to answer different types of questions

In isolated-question QA, several distinct, rigidly defined, question types can be identified. The most common are factoid, list, definition, and relationship questions (Voorhees, 2005). Other question types found in QA dialogue systems are analytic

(HITIQA) (Small et al, 2003), complex (a set of simpler questions, FERRET) (Hickl et al, 2006), encyclopedic (IMIX) and yes/no (IMIX). This chapter will mainly consider factoid and encyclopedic questions. Factoid questions are answerable by a single word or phrase, while encyclopedic questions are answerable by (slightly modified) text fragments, taken from a document out of a set of documents. The documents are typically a few paragraphs of text on a specific subject. In fact, the IMIX database documents are mostly sections from medical encyclopedias.

This classification concerns isolated questions only. Many or most QA systems assume that an FU is an FQ that can be handled as a regular factoid QA question. An FQ is usually limited to whatever you can feed into an IR engine that is made for isolated questions. Additionally, the FQ hardly ever refers to previous answers, because factoid answers are just single words and phrases. Consequently, FQ handling methods typically concentrate on how to adapt the input to the standard QA or IR process in order to include context from previous questions.

The dialogue tracks of TREC (Voorhees, 2005), QAC (Kato et al, 2004), and CLEF (Forner et al, 2009) have a particularly limited view of what an FQ is, in order to make evaluation easier. However, as a consequence, they suffer from other methodological problems, which was the reason why the TREC dialogue track was dropped (Voorhees, 2001). All FQ in the dialogue tracks assume that each question always has exactly one answer, consistent with the "factoid" paradigm. The dialogues then assume that the user never needs to react to these answers, but follows a planned path. In reality, answers may have varying degrees of correctness or completeness, and users will respond to this with a continuum from pre-planned FQ to FQ concerning details of the answer, to utterances that indicate the user is unhappy with the answer. For example:

- **Reactions to wrong or unclear answers.** Many answers will be wrong, and many encyclopedic answers will be only partial or unclear. Such bad answers should not cause the dialogue to disintegrate. The DAM should at the least know something about users' reactions when they are confronted with wrong or unclear answers.
- **Discourse questions.** These are questions that refer to the specific discourse of an answer (Theune et al, 2007). So, they cannot be understood without literally taking into account the answer.
- **Other meaningful FU that are not FQ.** Most QA research does not account for user reactions other than domain questions that are readily processable. In real-life dialogues, other phenomena are seen: utterances that indicate uncertainty about the correctness of the answer, negative feedback, and acknowledgements.

One goal of the Vidiam project was to explore what answering strategies are best for these kinds of FU. The examples in Table 1 illustrate some of the FQ it is felt a QA DAM should handle.

Table 1 Example utterances.

Typical TREC-style factoid follow-up question sequence:
Who is the filmmaker of the Titanic?
Who wrote the scenario?
How long did the filming take?
Possible non-factoid follow-up question sequence:
What do ribosomes produce? (factoid initial question)
All types of proteins?
And how does that work?
Some "real life" FQ from the corpora:
So the answer is "no"?
Shorter than what? Another form of schizophrenia?
How often is often?
What are these blue spots? (multimodal FQ)
Is this always true? (a reaction to an explanatory answer)
What does 'this' refer to? (reference to a pronoun in the answer)
One minute per what? (reply to a factoid answer "one minute")
What is meant here by 'hygienic handling'?

2 Overview of Existing Systems

This section provides an overview of existing interactive QA systems, that handle FU and are based on unstructured QA. In all systems, it is possible to distinguish a DAM and a QA engine. The DAM is the software that determines a question's context from previous utterances and gives responses other than QA answers. It is conceivable that this software is integrated in such a way that it cannot be practically separated from the QA engine, but in practice, it is usually quite clear how context is calculated and passed to a QA/IR system.

Previous work (van Schooten and op den Akker, 2005) distinguishes between a "black box" and an "open" QA engine, based on what information the DAM can pass to the QA engine. The black box model requires isolated plain-text questions. This gives the fewest possibilities but has the advantage of modularity. The black box model requires the DAM to rewrite any FQ into a self-contained question, which, as discovered during the IMIX project, is both difficult and conceptually problematic (van Schooten and op den Akker, 2005). Unfortunately, the QA engines in IMIX are all black box. So, the possibilities of open QA engines have been assessed by means of off-line analysis, and collaboration with the Ritel system, which has an open QA engine.

If there is an open QA, the ability to pass context depends on the nature of the QA. Typical QA systems have a separate information retrieval (IR) stage, but there are differences between individual systems. In particular, the internal representation of the IR query varies. Some QA systems translate a question into an IR query by classifying it into a specific question type with appropriate arguments, others build a more complex semantic frame from the question, such as SmartWeb (Reithinger et al, 2005) and Ritel (Galibert et al, 2005). The IR itself may be done in different ways, such as using semantic document tags resulting from a pre-analysis phase

Table 2 Overview of existing QA systems that handle FU. The **discourse** column indicates that a system handles discourse questions. The **domain** column gives a coarse characterisation of the domain, with *open* being an open-domain system in the classical sense. *GUI* indicates that the system uses graphical user interface style interaction for FQ handling. *Qtype* indicates that the question type (that is, person, date, number, etc.) may be passed as dialogue context to the IR. *Keyword* or *kw* indicates that keywords or key phrases may be passed as dialogue context. *Blackbox* indicates that only full NL questions may be passed to the IR.

System	Language	Speech	Multimodal	Discourse	Question types	QA-DAM interface	Domain
IMIX	Dutch	-	Yes	Yes	Encyclop.	Blackbox	Medical
Ritel (Galibert et al, 2005)	French	Yes	-	-	Factoid	Qtype+kw	Open
Hitiqa (Small et al, 2003)	English	-	-	GUI	Analytic	GUI	News
De Boni et al.'s system (De Boni and Manandhar, 2004)	English	-	-	-	Factoid	N/A	Open
SmartWeb (Reithinger et al, 2005)	German	Yes	Yes	GUI	Factoid	GUI	Multiple
Rits-QA (Fukumoto et al, 2004)	Japan.	-	-	-	Factoid	Blackbox	Open
ASKA (Nara institute) (Inui et al, 2003)	Japan.	-	-	-	Factoid	Qtype+kw	Address
KAIST (Oh et al, 2001)	English	-	-	-	Factoid	Qtype+kw	Open
NTU (Lin and Chen, 2001)	English	-	-	-	Factoid	Keyword	Open
OGI school's system (Yang et al, 2006)	English	-	-	-	Factoid	Keyword	Multiple
FERRET (Hickl et al, 2006)	English	-	-	GUI	Complex	GUI	News

(such as IMIX's Factmine and Rolaquad), or by matching syntactic structures of question and answer (such as IMIX's QADR). Because of all these variations, it is assumed that a DAM will only be able to give relatively simple and general hints to the QA back-end, such as pointers to the questions that should be considered context for the current question.

Table 2 attempts to give an exhaustive list of QA systems that handle FU from the literature, and compares these systems according to the features identified.

2.1 FQ Context Completion Strategies

Depending on the QA-DAM interface, passing the required context to the QA for answering a regular FQ can be done in several ways. In current QA dialogue literature, it is possible to distinguish three basic approaches to handling FQ:

1. **Rewriting a follow-up question to a self-contained question.** This is applicable to black-box QA. The effectiveness of a rewritten sentence still depends on the internals of the QA engine. An advantage of rewriting is that a successfully rewritten question ensures that the interpretation of the FQ is correct and

complete. Moreover, this correctness and completeness can be readily evaluated by a human annotator by simply judging whether the question is self-contained and answerable. A sentence can be rewritten in different ways, and according to different criteria, such as:

a. all appropriate search terms occur in the rewritten sentence
b. sentence is syntactically and semantically correct
c. sentence is as simple as possible
d. sentence is answerable by a human "QA"
e. the QA gives the correct answer

The criteria that are emphasised here are (b), (c), and (d). (a) is believed to have been insufficient given the number of non-search-term approaches, and is contained in (d). While empirical purists may consider (e) to be the "ultimate proof" of suitability, in reality, it depends on both the quality of the particular QA and the document database used. The criteria (b)-(d) have the additional advantage that they can be evaluated readily by a human annotator.
Basic forms of rewriting include replacing an anaphor with a description of its referent, and adding missing phrases to elliptical or otherwise incomplete sentences. Such rewriting satisfies criteria (a)-(d). For example, the (Japanese) Rits-QA system (Fukumoto et al, 2004; Fukumoto, 2006) uses two kinds of ellipsis expansion, and anaphor resolution. Their scheme managed to rewrite 37% of the FQ in their corpus correctly.

2. **Combining the FQ's IR query with that of previous utterances.** This is called the *IR context* approach. This approach requires an open QA engine. The advantage is that it can take short cuts, which may mean avoiding having to fully interpret an FQ when not strictly necessary. As an example of this approach, consider the (Japanese) Nara Institute system (Inui et al, 2003). It handles FQ using the IR context, based on its "question type/keyword" based IR:

   a. Analysing the question, obtaining keywords, question type, and answer class. Obtaining question type and answer class from dialogue history if missing.
   b. Add keywords from dialogue history (from both system and user utterances).
   c. Remove keywords with low weights, or with the same semantic class as answer class. For each semantic class, keep only the keyword with the highest weight.
   d. If no answer is found, relax the current request until an answer is found.

   Note that step (b)-(c) are similar to a regular salience-based linguistic reference resolution scheme, although they avoid having to resolve specific referents. The (English) KAIST system (Oh et al, 2001) uses an approach similar to step (b)-(c), except that their reference resolution does resolve to a specific referent. Similar schemes are found in the (French) Ritel (Galibert et al, 2005) and (German) SmartWeb (Reithinger et al, 2005) system, both of which use merging of semantic frame representations. Ritel also uses request relaxation.

3. **Searching within previous search results.** Searching within the top n documents retrieved by a previous question seems to be a successful strategy, as pointed out by De Boni (De Boni and Manandhar, 2004). De Boni found that about 50% of FQ could be answered using this strategy (given a perfect enough QA). Such a facility would even enable users to use a strategy of incremental refinement, that is, using multiple simple questions to assemble a query throughout multiple utterances.
   A simple version of this method is searching only the document where the previous answer came from. This has some significant advantages in terms of simplicity. In particular it is a method that will work for any QA engine that works by selecting text fragments from documents.

2.1.1 Universal steps in the context completion process

It was found that context completion follows the same basic strategy in different systems. Three main steps can be distinguished, each of which can be developed and evaluated separately. It can be argued that following this three-step approach explicitly will lead towards a more structured development methodology. QA contests may even design new tracks based on these steps.

1. **Identification of need for context completion.** This is analogous to detecting if the question is self-contained or not. This is the most basic step in any QA that wishes to handle FQ. Some QAs only implement this step, then apply some very basic IR algorithm, with good results. Note that existing TREC/QAC context tracks do not address this step at all, since the TREC/QAC dialogues do not contain topic shifts.
   In case a system has support for discourse questions or other FU, they can be detected in this step, as IMIX does.
   In the corpora many questions were found where adding context is not required, but not harmful either. Let this be, defined this a little more clearly. At one end there is the *harmfulness* condition, which indicates that adding any context is harmful (such as indicated by the classical "topic shifts"). On the other end, there is the *insufficiency* condition, where *not* adding context makes the question insufficiently complete to answer. In between, are basically the class of self-contained on-topic FQ.
   Features typically used to perform this step are: presence of pronouns and anaphora, ellipsis, general cue words, presence of keywords specific enough for successful IR, semantic distance of keywords with those of previous utterances. Performance is measured by the percentage of correct (neither harmful nor insufficient) classifications. Performance baseline is choosing the most often occurring FU class (which is typically the non-self-contained FQ).
   Ritel uses the notion of *topic shift*, which is meant to indicate that it is harmful to use the context completion machinery when the user has changed to a completely different topic. Ritel sets a context completion prohibition flag if

topic change is detected. Topic shift is also used by the De Boni and OGI systems. In fact, these two algorithms are based on detecting the boundaries of concatenated sequences of unrelated dialogues or TREC FQ series.
IMIX, on the other hand, tries to make a distinction between questions on the same topic that do and do not require context, even if they are follow-up questions. IMIX is trying to be as lazy as possible as regards context completion, because its particular algorithm is relatively error-prone. It is primarily based on distinguishing between different kinds of FQ in the FQ corpora, in which there are no topic shifts. The paradigm used here is basically the insufficiency condition.

2. **Identification of rewriting strategy and anaphora.** When trying to rewrite the question into a self-contained question, it is important to find out how it should be rewritten. Some systems, that do not require rewriting to obtain proper input for the IR, may still require some structural properties of the question to be passed to the IR, in which case this step must be performed partially. If the question is not being rewritten, but just passing "bag of words" information directly to the IR engine, as in Ritel, this step can be skipped entirely.
This step is explained in detail by van Schooten and op den Akker (2005). For each FQ, one of a small set of relatively basic rewriting strategies was chosen. The most successful strategy by far, for both unimodal and multimodal FQ, was found to be the anaphora substitution strategy. Here, an anaphor in the sentence has to be located and identified as being substitutable. In practice, it was found that only a fraction of typical FQ can be rewritten using this method, due to the lack of proper rewriting methods to cover the entire range of FQ satisfactorily.
Performance is measured by assuming that the first step was done correctly, and by counting the percentage of correct (intelligible and syntactically correct) sentences. Any simple baseline is likely to come up with very low performance, so a baseline to compare with is not needed.
3. **Referent selection.** Both rewriting and IR context completion require referents from previous utterances to be identified and selected, while the answer document set strategy does not. Each question and answer is scanned for potential referents, which are stored in the referent database. For multimodal answers, the referents include the pictures and the visual elements or text labels within the pictures. Unlike the other referents in the database, these are retained only as long as the answer remains on screen. Referent selection then amounts to the selection of suitable referents from referents previously entered into the referent database.
For multimodal utterances, picture annotations and gesture recognition is necessary as well to perform this step.
Typical features used in this step are: semantic matching, general importance of keywords, antecedent salience, anaphor-antecedent matching, confidence, presence of deictic gestures. IMIX implements all of these except semantic matching.
For multimodal referent selection, gesture referent recognition is also required (recognising what visual element the user is pointing at), which in turn requires

Table 3 Context completion steps performed by different systems, and some performance figures. *"Yes"* means the step is performed but no figures are known; Baseline performance scores are shown between brackets. The *need-context* step is the most commonly evaluated one, and performance of different systems could be compared, except that the corpora are very different. What is measured in *overall* performance evaluations varies between systems (see the footnotes for what is evaluated), and cannot meaningfully be compared.

System	Need-context	Rewriting	Ref-select	Overall
Ritel	Yes	-	Yes	
IMIX	75%(61%)(1)	Yes	Yes	14%(2)
De Boni	83%(62%)(3); 96%(78%)(4)	-	-	N/A
Rits-QA	-	Yes	Yes	37%(6)
Nara	-	-	Yes	100%(7)
KAIST	-	Yes	Yes	(8)
NTU	-	-	Yes	(8)
OGI	93%(62%)(3); 74%(64%)(5)	-	Yes	84%(9)

(1) - unimodal FQ corpus, classification includes discourse question
(2) - unimodal FQ corpus, rewriting and ref-select combined
(3) - sequence of TREC-10 context dialogues
(4) - De Boni dialogue corpus
(5) - HANDQA dialogue corpus
(6) - QAC2 corpus, overall rewriting performance
(7) - overall context completion performance using restricted language dialogue
(8) - TREC-10 context task participants, no results
(9) - retrieval performance in top 50 documents, TREC-10 and TREC 2004

pictures shown as part of answers to be annotated with the location and name of the visual elements they contain.
Performance is measured by assuming that previous steps were performed correctly, and counting the number of cases in which no harmful nor insufficient antecedents were selected. It is argued that a baseline is applicable here. Selecting the most important keywords (such as all named entities) from the previous question has a relatively high chance of success. In fact, the OGI system uses this baseline method with success (Yang et al, 2006). Additional evidence is that the IMIX follow-up question corpus also shows that 75% of the anaphoric references refer to a domain-specific noun phrase in the original question. Bertomeu et al (2006) also found that 53% of all FQ in their dialogues refer to a topic introduced in the previous question.

Table 3 summarises existing systems in terms of these steps, and give performance figures where available. It has to be concluded that the performances are hard to compare. Not only do the systems have different languages and domains, they also use different performance criteria and corpora. It was found that the same system tested on different corpora can give quite different results. It was also found that no distinction was made anywhere between harmful and insufficient.

3 The Corpora

This section describes the three corpora collected and analysed. The first two are the follow-up question corpus (van Schooten and op den Akker, 2005), composed of 575 text FQ, and the multimodal follow-up question corpus (van Schooten and op den Akker, 2007), composed of 202 multimodal (text + gestures) FQ to multimodal (pictures+text) answers. Both are Dutch-language corpora. These are based on presenting users with canned questions and answers, which can be responded to by uttering an FQ. The third corpus is collected from user dialogues with the Ritel system, the French-language Ritel corpus (van Schooten et al, 2007).

A special method for collecting the Vidiam corpora was used, which is low-cost and specifically tailored for FU in QA systems. Instead of having an actual dialogue with the user (using either a dialogue system or a Wizard of Oz set-up), the user reacts to a set of canned question/answer pairs. The first user turn consists not of posing a free-form question, but of selecting a question out of an appropriately large set of interesting questions. The second turn of the user consists of posing a FU to the answer then presented. The dialogue simply ends when the user posed his/her FU, and there is no need for the system to reply to the FU, hence there is no dependency of the corpus on any dialogue strategies used. An obvious weakness of the method is that the dialogues are only two turns long. However, it was found that such a "second dialogue utterance" corpus can be rapidly collected, and contains many or most of the most important phenomena that a first version of a dialogue system will need to support. The first conclusion is that this is a very good low-cost corpus collection method for bootstrapping a QA dialogue system.

3.1 The Follow-up Question Corpus

The first corpus collected is a "second utterance" corpus consisting of text-only dialogue (van Schooten and op den Akker, 2005). A collection of 120 hand-picked questions with selected answers was first created. The collection was chosen so as to cover a variety of different question types and answer types and was also used to evaluate the IMIX QA engines. Answer size ranges from a single phrase to a paragraph of text. The answers had a proportion of fully or mostly correct answers (93 questions, the answers were retrieved manually), a proportion of wrong answers (20 questions, the answers are real output from the QA system), and a proportion of "no answer found" (7 questions). The users participated in the experiment through a Web interface. First, they had to select at least 12 questions which they found particularly interesting. Then the answers were displayed. For each question-answer pair, they had to enter a FU that they thought would help further serve their information need, imagining they were in a real human-computer dialogue. 100 users were asked to participate, who were mainly people from the computer science department and people working in a medical environment. 575 FU from 40 users were collected. The questions chosen by the users were reasonably evenly distributed. Almost all questions were chosen between 1 and 10 times by users; there was no question that was not chosen by any user.

Examining the corpus, it was soon found that the FU could meaningfully be classified into a number of distinct classes. The corpus with these classes were annotated.

Three main classes of FU (See Figure 1) were found:

- **Follow-up question (56%)**. All domain questions were considered that should be understood in the context of previous utterances, and which can meaningfully be interpreted literally, to be FQ. They illustrate that the user acknowledged the answer at least partially, and indicate a further user information need. Some of the FQ in the corpus contained cue words indicating their "follow-up" nature, but there were none with politeness forms or indirect forms. A significant part of these were effectively self-contained, even though they were clearly on the same topic (25% of all FQ).
- **Negative feedback (28%).** This includes negative feedback questions and statements, questions indicating uncertainty about correctness of the answer, and reformulations indicating the user was not happy with the answer. Several distinct types of these were found:

  1. **Negative questions and statements 20%.** There seemed to be two main forms of these: repetition of the original question, with or without negative cue phrases (with no serious attempts at reformulations made); and a negative remark, usually simple but sometimes containing corrective information. In general, it appeared that there was relatively little useful information to be obtained by further analysing the negative feedback utterances. In some cases, a negative question and statement were combined in a single utterance.

q:	wat zijn hartkloppingen?	*What are heart palpitations?*
a:	De patiënt neemt dit waar als hartkloppingen.	*The patient experiences this as heart palpitations*
fuu:	Maar wat zijn hartkloppingen dan?	*But what are heart palpitations?*

Repetition example

  2. **Verify questions 3.6%.** Questions that indicate that the user is not sure about the meaningfulness or correctness of the answer.

q: Hoe merk je dat je hoge bloeddruk hebt?
What do you notice when you have high blood pressure?
a: Hoge bloeddruk (hypertensie) is meestal een aandoening die over het algemeen geen symptomen veroorzaakt.
High blood pressure (hypertension) is usually an affliction that generally does not cause any symptoms
fuu: Dus je merkt niet dat je een hoge bloeddruk hebt?
So you don't notice anything when you have high blood pressure?

Verify question example

  3. **Reformulations (4.4%).** These usually occurred when the system gave a "no answer" response. They are generally self-contained questions without any special linguistic cues.

q: Komt RSI in Nederland vaker voor dan in de rest van Europa?
Does RSI occur more often in the Netherlands than in the rest of Europe?
a: RSI komt niet alleen bij beeldschermwerkers voor maar ook in de industrie en bouwsector.
RSI does not only occur among screen workers but also in industry and construction.
fuu: hoe vaak komt RSI voor in nederland vergeleken met de rest van europa
how often does RSI occur in the nederlands as compared to the rest of europe?
Reformulation example

- **Acknowledgements (13%).** Almost all acknowledgements consisted of a one- or two-word acknowledge phrase, such as "ok" or "thanks".

For the rest of the article, the above classes are named resp. **FQ**, **negative**, **verify-question**, **reformulation**, and **acknowledge**. Everything else was labelled as **other**; this covered only 2.8% of the utterances. About half of these could be classified as "meta" requests, such as asking for a literature source, or requests concerning search strategy or answer form.

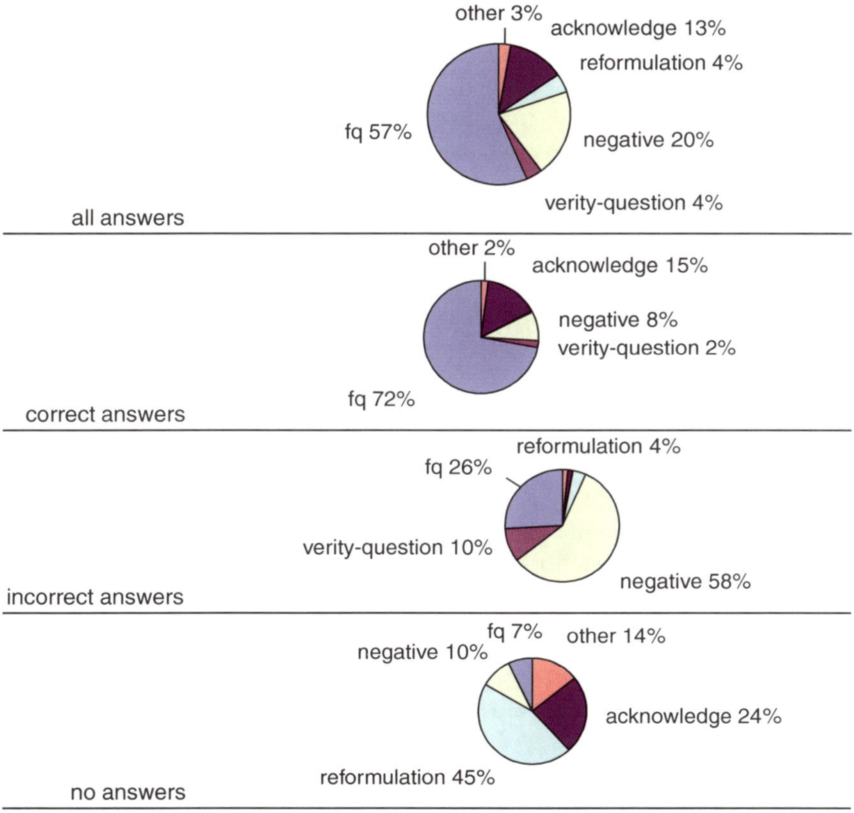

Fig. 1 Types of FU in corpus. FQ = follow-up question. self-contained = follow-up question that can be understood without context. negative = negative feedback.

In order to get an idea how successful a rewriting strategy would be, the FQ portion (56%) was split into classes, based on how they refer to the dialogue context. In particular, to find out if FQ could be rewritten, an attempt was made to rewrite each FQ manually into a self-contained question, satisfying rewriting criteria (b)-(d). Several special subclasses of rewritable FQ were identified, based on the existence of machine-ready transformations that can be used to rewrite them. The following transformations were identified:

- **anaphor**: FQ with anaphora which refer to NP antecedents,
- **elliptic**: elliptic FQ (FQ without verb) which could be expanded by including constituents from a previous sentence,
- **other-pp**: FQ which could be expanded by attaching a PP at the end, consisting of a prep and an NP from a previous utterance, or which is a PP from a previous utterance.

The rewritable questions that did not fall into these categories were labelled as **referencing-other**. The estimation is that most of these will be difficult to rewrite without rich knowledge. Some FQ did not require rewriting, even though almost all of them are clearly within the context of the previous question or answer in terms of information need. These were labelled these as **self-contained**.

One more surprising class emerged: namely, a significant number of FQ turned out to be in fact demands of the user to show a missing part of the text that the text fragment (implicitly or explicitly) referred to. It was not possible to rewrite these FQ, except by actually quoting (parts of) the answer literally. These are labelled as the **missing-referent**. An example:

q:	Waar duiden koude voeten op?	*What do cold feet indicate?*
a:	Hierdoor kan een verhoogde bloeddruk ontstaan of een vernauwing van de bloedvaten in het been.	*This can cause heightened blood pressure or a constriction of the blood vessels in the leg.*
fuu:	waardoor?	*What can?*

Missing referent example.

Figure 2 shows the breakup of FQ into the different classes. To evaluate the validity of the classification, an inter-annotator agreement analysis was also performed, which resulted in a Kappa (Cohen, 1960) of 0.62 over 11 utterance classes in total. This is not a very high value, but it is considered sufficient considering the large number of classes.

3.2 The Multimodal Follow-up Question Corpus

In order to get some data on how users would react multimodally to a multimodal answer, a corpus of multimodal follow-up questions was also collected. It was again collected by means of a set of 20 canned questions with multimodal answers, from which the users select 10. The user is then presented with the answer to each question, to which the user can pose a free-form multimodal follow-up question. The

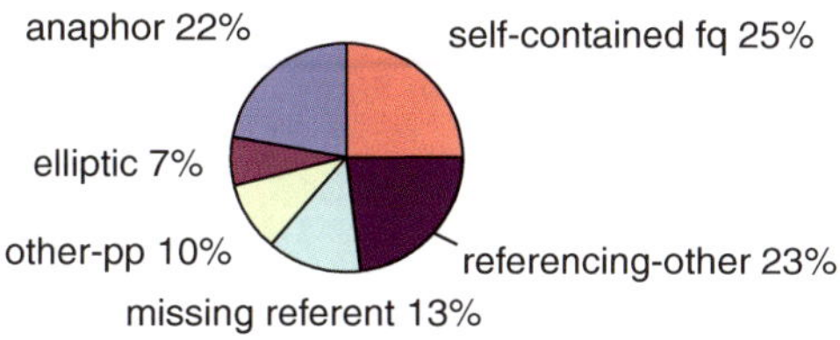

Fig. 2 Types of FQ in corpus. anaphor = question contains anaphor referring to NP, elliptic = question is elliptic (no verb) and can be expanded by adding constituents from a previous sentence, other-pp = can be rewritten by adding PP with NP from previous sentence, missing-referent = FQ that requested for something missing in the text fragment, referencing-other = all FQ that could be rewritten, but not with any machine-ready technique, self-contained = FQ which can be understood without context.

users were asked to pose multimodal follow-up questions, rather than any kind of follow-up question, in order to arrive at an appropriately large number of multimodal utterances for analysis. "Multimodal" was primarily defined as referring to pictures in the answer, with or without using pointing gestures. This has the disadvantage of artificiality, which means that the results may be biased in unknown ways, which is not a fatal flaw, since the range of phenomena that may occur is primarily of interest, rather than their precise relative frequencies.

The users' output modalities were typed text and pointing/gesturing with the mouse. The users were computer science students and employees, which could access the experiment through a Web page. Before the start of the experiment, the users were presented with one page of instructions, several examples, plus a short introduction to the user interface. To make the multimodal aspect of the answers significant, particularly interesting and complex medical pictures and diagrams were chosen for the 20 question/answer pairs. 202 multimodal "second questions" from 20 users were collected. Figure 3 gives an impression of the collected data used in this research. Note that the text is in Dutch.

3.2.1 Utterance type and possible utterance rewriting

In multimodal QA, it was found that "non-QA" questions occur more often than in unimodal QA. The most common type is asking for the identity of a visual element. That is, users say something like "What is this?", or "Is this the .....?" while indicating a visual element. Other kinds of visual discourse related questions also occur, for example, "Of what side of the head is this picture?"; "Where in the picture is the medicine?"; "In what direction do these flows go?". Following this line of thought, the follow-up questions in the corpus were classified into different types.

It was found that 19% of the questions were not multimodal (mostly regular questions that did not include mouse pointing and did not refer to any visual referent), or were not follow-up questions (these are mostly remarks). There are

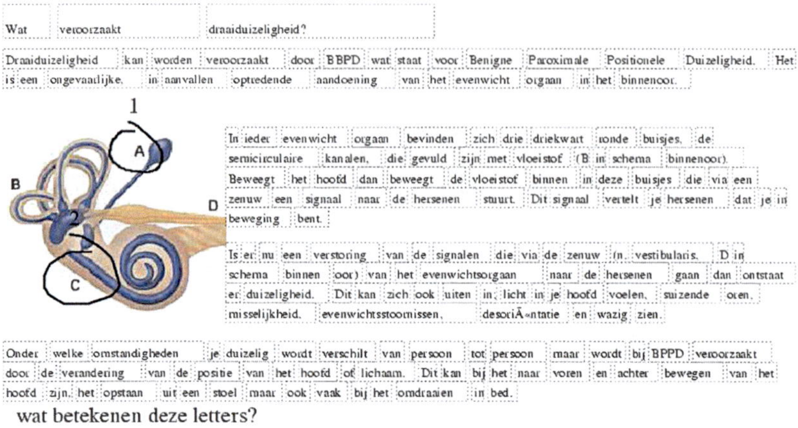

Fig. 3 Example interaction from the corpus. At the top is the original question, in the middle is the answer, at the bottom is the user's follow-up question. The encirclings in black are the mouse strokes made by the user. The stippled boxes are the words' bounding boxes. Original question: "What causes vertigo?", follow-up question: "What do these letters mean?".

discards in the results presented here. Everything else was classifiable into five classes (see also Figure 4):

- **Self-contained (18%)**. The question does not need any rewriting to make it answerable by a QA.
- **Deictic-anaphor (22%)**. A regular QA question, which is rewritable using anaphor substitution to form a self-contained question. A DAM may handle this kind of question by detecting which transformation is applicable and finding the appropriate referents and inserting them.
- **Deictic-other-rewritable (13%)**. A question that can be (manually) reformulated so as to form a self-contained QA question. While not rewritable like regular-rewritable, these questions can be handled by a QA in the regular manner.
- **Visual-element-identity (20%)**. Not answerable without relating to the answer's specific discourse, but answerable by just naming the identity of a particular visual element of a picture in the last answer.
- **Visual-property (26%)**. Not answerable without relating to the answer's specific discourse, and has something to do with the content of a particular picture, other than visual-element-identity. This is a difficult type of question to handle properly, but might simply be answered by producing a suitable figure caption paraphrasing the entire figure.

3.2.2 Visual referents

In the corpus, users never asked a follow-up question by just pointing. There was always some text. In fact, almost all follow-up questions can be considered primarily

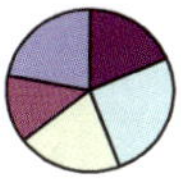

Fig. 4 Pie chart showing the percentage distribution of the follow-up question classes in the corpus.

linguistic, and the meaning of pointing within the interaction can generally be understood as just hints (though sometimes essential ones) to disambiguate the anaphora and other references in the question text. Referents in the utterances were usually visual elements, but some of the utterances referring to visual elements did not include pointing actions. Instead, the visual elements were typically referred to by means of their colour, shape, or name. Often, a redundant combination of these were used; for example, one user asked "What function do these blue spots have?" while encircling several blue circles, thus combining colour, shape, and pointing action as hints to disambiguate the utterance's referents. Overall, the findings indicate that traditional (anaphor) reference resolution is a meaningful and important first step in the interpretation of the multimodal utterances. The rest of this section tries to find out in what ways users refer to visual referents.

How do users indicate using the mouse? The use of encircling by producing several encircling examples was encouraged before the users started the experiment. Encircling the most important type of indication was considered, because it allows indication of both location and size of a visual element. However, as is usual in "natural" dialogue systems, the users are allowed to use any other kind of pointing action, and users did commonly use several other types of pointing action.

In order to provide a more systematic analysis, three aspects of the user utterances were looked at: the pointing gestures, the anaphora, and the possible relations between these two.

The first aspect involves segmenting the series of mouse strokes of one utterance into a set of pointing gestures. A mouse stroke is defined as a continuous line or curve drawn with the mouse, and a pointing gesture as a set of mouse strokes that has the goal of indicating a visual element. It was found that most pointing gestures consist of only one mouse stroke, but some, like arrows, typically consist of two or three mouse strokes. Almost all mouse strokes were found to be clearly identifiable as being part of pointing gestures. 81% of the utterances were found to contain at least one pointing gesture, and 23% of these (19% of all utterances) contained multiple ones.

A type was assigned to each pointing gesture. Four distinct types were found to be sufficient to cover almost all cases: encircle, tap (that is, just a mouse click), underline (as in, underlining a word), and arrow. What was left were a small number of mouse strokes that appear to be erroneous (labelled as error), and a small miscellaneous category (labelled other). Figure 5 shows the relative frequencies of the different classes.

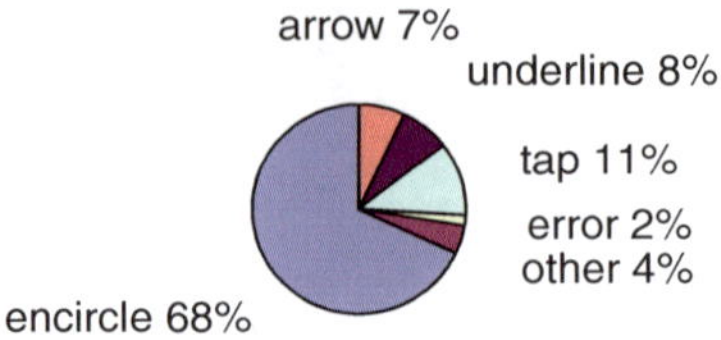

Fig. 5 Pie chart showing the percentage distribution of pointing gesture types.

The second aspect looked at is the relation between the pointing gestures and the linguistic components of the questions. This was done by labelling the anaphora that clearly referred to visual elements, and the anaphora that clearly correspond with pointing gestures. Only a small minority of pointing gestures (9%) were discovered, for which no anaphor could be pointed out, indicating that anaphora and pointing gestures are closely related. No gestures corresponding to multiple anaphora were discovered, but some anaphora did refer to multiple gestures. It was found that these anaphora were significantly more frequent in plural form, explicitly indicating a set of referents. However, both singular and plural forms were found in all cases (see Figure 6).

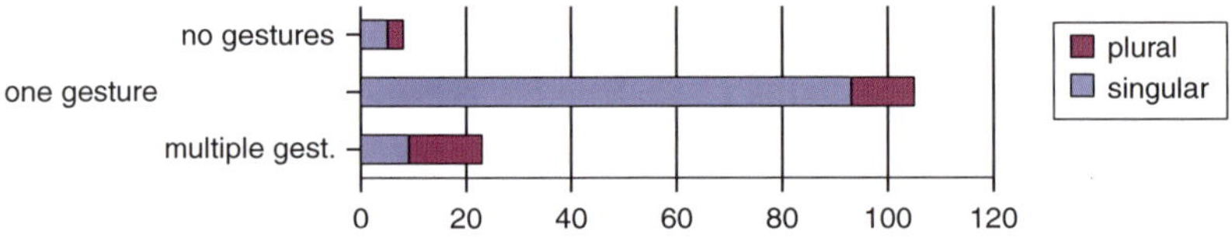

Fig. 6 Bar chart showing the anaphor-gesture relationship: the number of gestures that each anaphor corresponded with (either none, one, or multiple), and the plurality of the anaphor.

The third aspect that was annotated is the ways in which the identified anaphora provide hints towards resolving their referents, beside the pointing gestures. Hints are classified according to the aforementioned types: colour, shape, and name. 49% of the anaphora were found to provide no hints, they were just indications like "this" or "this area". Of the remaining 51%, name occurred the most often by far. Colour and shape were relatively often used simultaneously. As one might expect, name was almost never used simultaneously with colour and shape. The findings are summarised in figure 7.

3.3 The Ritel Corpus

The Ritel corpus is the only corpus collected from a fully functional dialogue system. The Ritel system (Galibert et al, 2005; van Schooten et al, 2007) is a factoid open-domain QA dialogue system that works over the phone. Users can call Ritel and ask questions using free-form speech utterances. The system answers with a

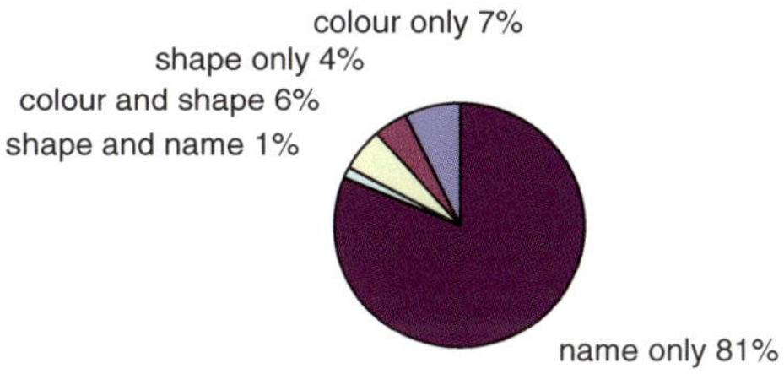

Fig. 7 A pie chart showing the relative distribution of hints used in anaphora.

factoid type answer, which is usually just a single word. Dialogue management functionality was added, which mainly concerned confirmation whether the recognised keywords, question type, and answer were correct, and as to signalled incomplete or non-understanding.

Users were invited to call Ritel over the phone. 15 dialogues in this manner were collected. These comprise a total of 244 user utterances. These were annotated with dialogue act type, presence of topic shifts, and self-containedness, mostly following van Schooten and op den Akker (2005); see Table 4. A vast majority of the utterances were questions. The non-questions were mostly explicit topic and type mentions (such as *"This is a new question"* or *"I have a question about ..."*, and answers to system questions. There were only a few explicit disconfirms, all of them by one user.

How well did the ASR manage to recognise the relevant information in the different types of utterance? To measure this, the ASR results were subdivided according to whether the essential information was recognised correctly. 131 utterances (54%) were found to be sufficiently well recognised, that is, all relevant keywords and answer type cues were recognised, as well as any relevant communication management cue phrases. Some 76 (31%) were partially recognised, that is, at least part of the IR material and dialogue cues. This leaves 37 utterances (15%) which were not recognised to any useful degree.

Some user act types were found where the ASR performance distribution deviates significantly from the overall one. Relatively well recognised were topic announcements, negative feedback, and non-self-contained FQ. Particularly poorly recognised were reformulations, self-contained FQ and repetitions. This seems to be related to the presence of domain-specific keywords, such as named entities, which were the toughest for the ASR. Interesting here is that non-self-contained FQ were better recognised than self-contained ones because, typically, the named entities were left out. This suggests that context completion can be useful if it has already established the most difficult keywords earlier in the dialogue.

To further examine the dialogue interactions between user and system, the subdivision of the different system utterance types were looked at, the user's feedback utterances, and the relationships between them. There were 229 system responses in total, subdivided as in Table 5. Most user feedback is implicit, consisting of informs (users giving partial information in a non-question form, mostly responding to a system question), and repetitions and reformulations. A

Table 4 User dialogue act types found in the corpus.

29	(12%)	new questions (that is, in-dialogue topic shifts)
74	(30%)	FQ (27 non-self-contained, 47 self-contained)
87	(36%)	reformulations, repetitions and informs
18	(7%)	negative feedback or topic announcements
7	(3%)	byes
12	(5%)	miscellaneous utterances

Table 5 Types of system responses in the corpus.

115	(50%)	give an answer and confirm IR material
55	(24%)	confirm answer type and ask for more info
43	(19%)	confirm keywords and ask for more info
7	(3%)	ask the user to repeat
9	(4%)	indicate non-understanding

Table 6 Types of user feedback in the corpus.

47	repetitions (of which 35, or 74% were self-contained; 60% were after system asked for more info)
23	reformulations (almost all were self-contained; 52% were after system asked for more info)
17	informs (almost all were partial, most (75%) were after system asked for more info)
12	topic change announcements
6	explicit topic announcements
3	disconfirms (all by one user)

minority were direct negative responses or explicit announcements of a new topic, see Table 6.

So, it was found that almost all corrections are repetitions or informs. As far as the confirmation strategy is concerned, it appears that confirmation was not picked up in the sense that users confirmed or disconfirmed anything explicitly, but users did use it to determine whether they were dissatisfied with the response. What users mostly did was essentially repeat when they found the system reply unsatisfactory, which means that the "repeat at least twice" kind of confirmation will work well.

If only the difference between when the user repeats or when the user poses a new question can be detected, this can be properly used to handle confirmation. However, it is not clear how to do this. Most repetitions and reformulations have no or few surface cues to help detect them, although informs could be detected by looking at the absence of a question form.

The system was quite successful at detecting topic change announcements. This was less so for explicit topic/type mentions. While the system tags phrases that can be considered cues for topics, no significant correlations were found with topic announcements, topic shifts, or self-contained versus non-self-contained questions.

4 The Dialogue Manager

In this section, how the corpus results were used to design and evaluate effective dialogue strategies is described, based on the collection of strategies explained in Section 2. What is implemented in the Vidiam DAM is described, and how well the DAM performed. The overall approach is to implement simple, knowledge-poor techniques and depend as little as possible on a specific QA architecture. Three aspects of dialogue performance are described: FU classification, according to the classification scheme, context completion performance, and overall answering performance.

4.1 FU Classification Performance

The FU classifier was first implemented by manually selecting words, phrases, and POS tags, that would be cues for specific classes according to intuitive language theory, and which would give the best performance when tested on the corpus. This was then compared with machine learning classifiers, using all unigrams, bigrams, and trigrams of words and POS tags in the sentence as input features.

The manual approach has two advantages over machine learning: prevention of non-generalisable results or "overfitting" (since the corpus is really a bit small for machine learning), and obtaining insight into the language phenomena involved. If the manual classifier has a performance similar to an optimal machine learning algorithm that uses a wide choice of features, this gives an indication that the intuitive algorithm did not miss something important.

With the manual algorithm, a performance of 55% was obtained over all utterances in the corpus. The Weka toolkit (Witten and Frank, 2005) was used to try different machine learning algorithms. The best performance found was in the 55%-60% range, very close to the manual algorithm. The different algorithms and feature selection filters also tended to select mostly the same features as the manually chosen ones. The binary tree classifier even came up with binary decision trees similar to this one. A performance of 62%-63% was managed by using support vector machines, and adding an extra feature which denoted whether the semantic classifier found a sufficiently specific medical domain term in the sentence.

This is a reasonable result for such a knowledge-poor approach. It is enough to start with, especially since the manual algorithm concentrates on avoiding false positives with respect to non-default behaviour, rather than optimising overall classification performance per se. For example, misclassifying an "anaphor" FU as "negative" would be a disaster, while misclassifying "anaphor" as "referencing-other" only does limited damage. This ensures that most misclassifications are relatively safe, rather than costly, in terms of dialogue performance. In particular, referencing-other and self-contained were chosen as "sensible defaults" in case of uncertainty. If all matches of self-contained and referencing-other are considered as safe defaults for the utterances classified as FQ, the manual algorithm arrives at an accuracy of 73%.

In the rest of the section, the most important rules used for classification are described. These rules are very simple, and most of them seem transferable to other languages.

The words *niet* (*not*) and *geen* (*none*), together with the utterance not being a question, detected some 80% of negative statements. Negative questions could be detected using *maar* (*but*) at the beginning of a sentence or phrase, or the occurrence of *of niet* (*isn't it*).

Some 75% of acknowledgements could be detected by the words *dank* (*thank*) and variants, *ok*, and *oh*, *jammer* (*a pity*), *duidelijk* (*that's clear*), *mooi* (*nice*); and *dan* (*so*) at the beginning.

A variety of special cue words could be used to detect some subclasses of difficult-to-rewrite sentences (which should be classified as referencing-other). In the corpus *zo'n* (*such a*) (indicating reference based on similarity, which cannot be handled), *zoveel* (*that many*) (referring to quantities) can be found, *andere* (*other*) (referring to set operations).

Questions starting with the word *dus* (*so*) were found to be almost always verify-questions.

The occurrence of certain wh-words at the end of a sentence (indicating a question written in statement form) was a good indicator of some of the cases of missing-referent. In particular ... *wat?* (... *what?*), or .... *waarvan?* (... *of what?*), ... *waardoor?* (... *by what?*).

For analysing anaphora, the presence of regular determiners was looked at, like *de* (*the*) and *deze* (*this*), and PP-type determiners, which are a specifically Dutch phenomenon, for example, *erdoor* (*by it*), *hiermee* (*with this*), *daarvan* (*of that*). This detects most of the instances of the anaphor class along with the positions of the anaphora.

4.1.1 Multimodal FU classification performance

In the Vidiam DAM, multimodal support was added by extending the utterance type classifier with the utterance types found in the multimodal dialogue corpus. In particular, the visual-element-identity and visual-property types proposed in Section 2 are completely new types requiring special handling.

However, distinguishing between all unimodal and multimodal FU types was found to be a difficult task. All follow-up questions are fed from the corpus, including the non-multimodal ones, into a machine learning tool, using the number of mouse strokes, and the number of occurrences of specific words and part of speech tags in the sentences as features. The tool had to classify the combined set of classes identified for both unimodal and multimodal FU. Until now, it has not been possible to obtain a classification performance above 40%. The number of gestures in the utterance were found to be particularly important features (with no gestures indicating the question is non-multimodal, and multiple gestures, that the question concerns visual discourse), as were the occurrence of the determiner "this" (indicating a regular follow-up question with a deictic anaphor), and the occurrence of the word "picture" (indicating visual discourse). These simple rules were implemented into the Vidiam DAM. To obtain a better performance, high-level integrated knowledge, such as dialogue context is probably needed.

A gesture recogniser specifically for multimodal FU was also developed. It can handle taps, encircles and underlines. In addition to taps, encircle gestures were the most important gesture class. It was found that a simple bounding box algorithm, comparing the magnitude of the gesture's bounding box, the visual element's bounding box, and their intersection, could correctly identify 66% of all encircling gestures. A significant part of the failures concerned visual referents which were not annotated and are likely to not be naturally annotatable (16% of the gestures), and gestures encompassing multiple visual referents in one gesture (6%). Of the remaining referents, the algorithm managed to identify 88% of the referents correctly. This was considered a very good result for such a simple algorithm, showing that resolving gestures' referents is a relatively easy task. Taps, underlines, and arrows were also found to need different handling. Full support requires stroke segmentation and gesture type recognition, such as that found in Willems et al (2005).

4.2 Rewriting and Context Completion Performance

This section estimates what the corpus results mean for potential QA performance using different context completion approaches with the help of the (unimodal) follow-up utterance corpus. Regular FQ will be discussed here; other FU such as discourse questions will be addressed in Section 4.3.2.

Let's assume that the 13.2% of "missing referent" questions may be answered by just displaying more of the document the answer came from. It can be said that, of the 61.9% of FQ that are not already self-contained nor of type missing-referent, some 62.7% are potentially rewritable using relatively basic techniques (that is, they belong to the classes anaphor, elliptic, and other-pp). This is an *upper bound* for the rewriting approach, assuming that the user utterances can be correctly classified and rewritten.

The remaining ones may alternatively be resolved using the IR context approach. How many of these will be resolvable in this way cannot easily be determined, and depends on the inner workings of the IR engine and database. In contrast to the rewriting approach, it cannot be said with certainty if a certain IR context operation is "correct" and effective, given a specific QA engine.

4.2.1 Rewriting performance

Attempts were made to rewrite the anaphor, elliptic, and PP attachment classes of the unimodal FU corpus. Whether antecedents were found in the initial question or the answer was also looked at. Antecedents were often found in *both* the question and the answer. The following results were obtained:

Anaphor. Anaphor proved to be the least difficult to process, as the majority of these can be found using cue words and POS tags, while the antecedents can be found using a Lappin and Leass (Lappin and Leass, 1994) type salience algorithm. Some simplifications could be made. In particular, it was found that the antecedent could be found in the question in some 75% of the cases, so it was possible to optimise the salience algorithm by increasing the salience of antecedents found in the user utterance a little. Nevertheless, the achievements were limited. Of all FQ labelled by the system as "anaphor", only 42% were rewritten properly. This was in part due to a large number of false positives and in part due to errors in the reference resolution.

Elliptic. Elliptic proved to be more difficult. The classification performance was reasonable. 86% of all FQ of class elliptic was detected by just looking at the absence of a verb, with 44% false positives. Rewriting was done by finding the sentence that the elliptic expression referred to, and then using constituents from that sentence to form a self-contained sentence. However, performance of finding these antecedent sentences and rewriting were unusably low. Just a small minority of the antecedent sentences can be found by matching the elliptic sentence with words from previous sentences. There were no easy short cuts available either. Only the 55% of the elliptic FQ were found to refer exclusively to the original user question, and 14% referred to candidate sentences in both the user question and the system answer. Building a correct sentence from these proved difficult as well. Syntactic transformation using the Alpino dependency tree parser (Bouma et al, 2006) did not work in most cases. The first attempts showed that the transformed sentences were unsyntactic or did not make sense. Sentences were tried to be built from relation-argument type semantic frames using the Rolaquad semantic tagger, but the tagger did not seem reliable enough to get usable results.

Other-PP. The other-pp class of rewritable questions are relatively difficult to recognise, as FQ which require a PP to be attached are not readily distinguishable from self-contained questions. An obvious approach is to use semantic knowledge in the form of verb-argument relations, as is used in PP attachment disambiguation (Gildea and Palmer, 2001). Again, however, the available semantic tagger did not prove reliable enough for this. What was found is that 62% of other-pps referred to an NP occurring in both user and system utterance, and a total of 89% referred to an NP occurring in the user utterance. This suggests that the other-pp has at least potential for use in the IR context approach.

The first conclusion is that rewriting is not easy, and an accurate low-noise domain-specific semantic knowledge may be needed to do it. Such domain-specific techniques were not focused on here, so the current system only fully handles the anaphor class. Alternatively, anaphor and other-pp can be used as an IR context indicator that the user utterance should be used.

4.2.2 Potential Performance of the IR Context Approach

One of the most popular IR context approaches is to search only through the previous *n* documents retrieved by the previous question. De Boni and Manandhar (2004) reported a success rate of 50% for his own corpus (which is a combination of real-life and TREC dialogues). Why does this simple approach work so well? Intuitively, it is likely that most documents are coherent with respect to their topic, and that a single document is made so as to answer a number of common questions on that topic. In fact, this is underwritten by results from research by Lin et al (2003). They studied answer length in relation to user preference and behaviour. They found that, within a specific information need scenario, using longer text fragments as answers produced dialogues with less questions.

To gain more insight into this most basic IR context approach, the simplest version of it was evaluated, namely only looking at the document the answer came from (be it correct or incorrect). This way, the result depends only on the nature of these specific documents, and not on the way in which the IR matches the documents. The result obtained may be considered a lower bound for the performance of the IR context approach with respect to document selection performance, and an upper bound with respect to the expected document fragment selection performance.

It was checked manually whether the answer to each FQ in the unimodal FU corpus could be found in the document where the original answer came from. This was only for the *correct* answers, since considering incorrect answers here introduces noise related to the performance of the specific IR used. That is, the tendency of the actual IR will also be measured to either select the wrong document, or the wrong sentence from the right document.

All FQ were checked, including self-contained FQ. For about 1/3 of the answers, the source document could not be retrieved because of errors or incompleteness in their source references. The documents were nearly all sections from encyclopedias, and ranged from 50-500 words in size, with an average of about 150. A total of 196 FQ were checked this way. As a way of indicating the existence of vagueness in the documents' answering potentials, each FQ was annotated with a 3-point scale, thus including a "partial match" option:

- No match: the document could not answer the question in any way;
- Partial match: the document gave only a partial answer or an implicit answer;
- Full match: a satisfactory answer was explicitly stated in the document.

It was found that relatively few cases needed to be annotated as "partial match", so it was considered only the distinction "full match" versus "no full match". It was found that some 39% of the FQ could be answered by the document the answer came from. For the subclass missing-referent, this percentage was 73%. This high figure is consistent with the concept of missing-referent as directly referring to the document the answer came from. This means that FQ of this class can be effectively dealt with by directly showing more of the answer's source document.

For the remaining FQ classes, the percentage was 35% on average. Percentages for each class varied somewhat, ranging from about 20% to about 43%. The differences were not very significant, and in particular it was *not* found that self-contained FQ have a lower percentage of matches, in fact it was 43%. The lowest was elliptic, with 20% (3 out of 15).

This rather high figure may have some interesting implications. In fact, it is consistent with the stated intuition of document coherence being the real reason behind the success of this approach. Implementing a simple "use last document" strategy is likely to be worthwhile in any QA dialogue system, as long as there is a proper way to detect when the answer could not be found in the last document, and other strategies are available to complement it.

Another implication is that perhaps one should reconsider the ways in which to select a text fragment from a document. The results suggest that including more text in the answer text fragments may lead to better satisfaction of the users' information needs. On the other hand, the users seemed to prefer relatively small text fragments (as some started making comments on the text size when these were more than 4-5 sentences long).

4.3 Answering Performance

In this section the potential overall performance of a QA dialogue system is considered, based on the corpora.

4.3.1 Responding to non-FQ

In the unimodal FU corpus, it was found that 44% of the utterances are not FQ. Recognising and dealing with these classes of FU will already improve the system significantly. Even just reacting with appropriate prompts will help. In some cases, showing (more of) the document where the answer came from is a meaningful reaction. More sophisticated techniques can be imagined, involving system clarification questions for example.

Looking at the distributions for the different types of answer correct / incorrect / no-answer (see Figure 1), as might be expected, correct answers are replied mostly by FQ, and incorrect answers by negative feedback and verify-questions, although a significant minority were FQ. Almost half of the no-answers were spontaneously reacted to by reformulations, though the users were not prompted to do so. A quarter were acknowledgements, indicating the users accepted the absence of an answer. The "other" category was significant here. In fact, it was found that almost all "other" utterances amounted to explicit requests to search again. A dialogue system could easily react to this by an appropriate reformulation prompt.

The FU class seems to be an indication of whether an answer is satisfactory and/or correct. It is interesting to find out to what extent the quality of the answer can be discovered by just looking at the FU class. Significant is that no strong conclusions can be drawn when the user poses a FQ. But if the other FU classes' potential of classifying between correct and incorrect answers is looked at, the following patterns can be identified:

- Acknowledge almost always means correct;
- Verify-question almost always means incorrect;
- Reformulation almost always means incorrect;
- Negative usually means incorrect. There is a small but significant percentage of negative feedback to answers labelled as correct. A look at these answers indicated that the quality of these was less than that of most answers labelled as correct. This appears to be in part due to the fact that the answers were limited to selected text fragments from the corpus only. In these cases, a negative reaction is understandable. In fact, it is possible to use the user reactions found to reconsider the correctness of these answers in some cases.

These simple rules would allow to determine answer correctness to some degree. In particular, a large part of the incorrect answers can be predicted (72% to be precise, see Table 7).

Table 7 Prediction of correctness of answers by looking at FU type.

Actual↓ *Predicted*→	Correct	Incorrect	Unknown
Correct	15.3%	10.3%	74.4%
Incorrect	1.2%	71.8%	27.0%

4.3.2 Overall dialogue performance of the current system

In this section a technique to estimate the overall performance impact of the dialogue handling of the dialogue system offline is described. The unimodal FU corpus is used, and consider whether one has access to the simplest "search within previous results" method, namely, showing more of the document that the answer came from. The following two tasks were distinguished:

1. Identifying whether a FU is a FQ or not, and whether it is an acknowledge, negative feedback, or verify-question.
2. In case the FU is a FQ, passing the rewritten question and IR context hints to the QA, or producing the document in case the FQ is a missing-referent. For this task, two cases are considered:

a. QAs without any IR context abilities. It can be assumed that the DAM cannot pass IR context hints. The baseline is always passing the question as a self-contained question.
b. QAs with IR context abilities. It can be assumed that the DAM may pass IR context hints. The baseline is to consider all FU to be FQ, with only the context flag set.

With respect to these tasks, the cases in which the DAM would provide better, equal, or worse results than the baseline is considered, using the following criteria:

- **Better.** FQ is correctly rewritten; IR context is correctly specified; FU is correctly classified as negative, acknowledge, or verify-question; FQ is correctly classified as missing-referent.
- **Same.** Behaves the same as the baseline.
- **Worse.** FQ is wrongly rewritten while the original FQ would be a better query to the QA, or the wrong IR context hints are passed; wrongly identifies missing-referent; wrongly identifies negative, acknowledge, or verify-question.

The DAM has the following general strategy:

- If FU is negative-feedback, acknowledge, or verify-question, prompt accordingly (if present, it is possible to use any system clarification or answer selection strategies, currently there are none).
- If FU is missing-referent, show the document that the answer came from.
- If FU is another type of FQ, pass to the QA: the IR context (either the last question or the last answer), and the rewritten question. In particular:
  - anaphor: the IR context according to the predicted antecedent, and/or pass the rewritten question.
  - other: no specific hints, just pass that the question is a FQ.

For task 1 the following results were found:

Better	negative, acknowledge, verify-question identified	138 (24%)
Same	all FQ identified as FQ	405 (70%)
Worse	wrongly identified a non-FQ class	32 (5.6%)

For task 2 only the 405 utterances are considered classified as FQ. The following for case (a) was found (QA without context abilities):

Better	missing-referent correctly identified	8 (2.0%)
	question correctly rewritten	37 (9.1%)
Same	original question passed to QA	405 (70%)
	question wrongly rewritten, original not self-contained	52 (13%)
Worse	question wrongly rewritten, original was self-contained	1 (0.2%)
	incorrectly identified missing-referent	1 (0.2%)

The system does marginally better (11%) by grace of the missing-referent handling, the fact that the QA has no alternative to the anaphor rewriting, and the fact that there are hardly any false positives where it produces worse dialogue behaviour.

For case (b), the QA can handle context on its own, but using IR contexts shortcuts is an option, where these are possible. It was found that, in the anaphora, 89% which referred to a concept in the user utterance were detected correctly. As a strategy, it is possible to simply pass a pointer to the user utterance, instead of rewriting the sentence. This only gives a dubious improvement of 1%, however. The result would be:

Better	missing-referent correctly identified	8 (2.0%)
	IR context correctly specified	40 (10%)
Same	original question passed to QA	405 (70%)
Worse	IR context wrongly specified	49 (12%)

In other words, the number of cases which are worse is the same as the number which are better. Finally, looking at how much better the QA in case (b) handles FQ, assuming the IR context strategy discussed in Section 5.3, it was discovered that QAs are likely to find at least 35% of FQ by a default IR context approach, providing that the original answer is correct. For the non-self-contained FQ (75% of all FQ) this may give an improvement of 35%, which amounts to an improvement of up to 26%, depending on how many of the answers are correct.

5 Conclusions

This chapter examines dialogue and QA strategies in existing QA dialogue systems, and summarises the analysis of the different corpora that were collected. With help of the corpora the potential performance of a number of simple, knowledge-poor techniques for handling FU was assessed.

Although existing QA dialogue systems have different approaches to FQ handling and operate in different domains, it has been shown that they have certain tasks and a certain general architecture in common. In particular the following tasks were identified: identification of the need for context completion, question rewriting, and referent selection. We attempted to provide an account of the general issues to consider in these tasks. An interesting future direction is to develop a new set of standardised evaluations of these tasks, for example in the form of competition tracks.

In the corpus analyses, it was found that there is a broad range of ways in which users can express their information need during a QA session, in addition to regular FQ. It is also possible to conclude that a large part of these FU can be handled by a set of simple knowledge-poor techniques. However, there is still a large part of FU that cannot be handled this way, and mistakes made by the natural language processing system also result in a significant number of errors. The most difficult tasks are utterance type classification, and determining the exact context of an FQ. Only limited results were obtained using the range of techniques that were tried, and it is likely more knowledge-intensive techniques will be needed.

Multimodal FQ is still a new research area. It was found that it is possible to understand user pointing by interpreting pointing actions as hints for disambiguating linguistic anaphora. It has been shown how pointing can be handled using annotated pictures. The implementation remains infeasible, however, as long as there is a need to rely on labour intensive picture annotation techniques.

References

Bertomeu N, Uszkoreit H, Frank A, Krieger HU, Jörg B (2006) Contextual phenomena and thematic relations in database QA dialogues: results from a Wizard-of-Oz experiment. In: Workshop on Interactive Question Answering, HLT-NAACL 06, pp 1–8

Bouma G, Mur J, van Noord G, van der Plas L, Tiedemann J (2006) Question answering for dutch using dependency relations. In: Proceedings of the CLEF2005 workshop

Cohen J (1960) A coefficient of agreement for nominal scales. Educational and Psychological Measurement 20:37–46

De Boni M, Manandhar S (2004) Implementing clarification dialogues in open domain question answering. Journal of Natural Language Engineering

Forner P, Peñas, Agirre E, Alegrian I Forăscu C, Moreau N, Osenova P, Prokopidis P, Rocha P, Sacaleanu B, Sutcliffe R, Tjong Kim Sang E (2009) Overview of the clef 2008 multilingual question answering track. In: CLEF'08: Proceedings of the 9th Cross-language evaluation forum conference on Evaluating systems for multilingual and multimodal information access, Springer-Verlag, Berlin, Heidelberg, pp 262–295

Fukumoto J (2006) Answering questions of information access dialogue (iad) task using ellipsis handling of follow-up questions. In: Workshop on Interactive Question Answering, HLT-NAACL 06

Fukumoto J, Niwa T, Itoigawa M, Matsuda M (2004) RitsQA: List answer detection and context task with ellipses handling. In: Working notes of the Fourth NTCIR Workshop Meeting, pp 310–314

Galibert O, Illouz G, Rosset S (2005) Ritel: an open-domain, human-computer dialog system. In: Interspeech 2005, pp 909–912

Gildea D, Palmer M (2001) The necessity of parsing for predicate argument recognition. In: Proceedings of the 40th Annual Meeting on Association for C omputational Linguistics, Philadelphia, Annual Meeting of the ACL, URL `http://www.egr.msu.edu/~jchai/QAPapers/gildea-acl02.pdf`

Hickl A, Wang P, Lehmann J, Harabagiu SM (2006) FERRET: Interactive question-answering for real-world environments. In: ACL 2006, pp 25–28

Hofs D, Theune M, Op den Akker R (2010) Natural interaction with a virtual guide in a virtual environment: A multimodal dialogue system. Journal on Multimodal User Interfaces 3 (1-2):141–153

Inui K, Yamashita A, Matsumoto Y (2003) Dialogue management for language-based information seeking. In: Proc. First International Workshop on Language Understanding and Agents for Real World Interaction, pp 32–38

Kato T, Fukumoto J, Masui F (2004) Question answering challenge for information access dialogue – overview of NTCIR4 QAC2 subtask 3. In: Working notes of the Fourth NTCIR Workshop Meeting

Lappin S, Leass HJ (1994) An algorithm for pronominal anaphora resolution. Computational Linguistics 20(4):535–561, URL `citeseer.ist.psu.edu/lappin94algorithm.html`

Lin CJ, Chen HH (2001) Description of NTU system at TREC-10 QA track. In: TREC 10

Lin J, Quan D, Sinha V, Bakshi K, Huynh D, Katz B, Karger DR (2003) What makes a good answer? the role of context in question answering. In: Proceedings of the Ninth IFIP TC13 International Conference on Human-Computer Interaction (INTERACT-2003)

Martin JC, Buisine S, Pitel G, Bernsen NO (2006) Fusion of children's speech and 2D gestures when conversing with 3D characters. Special issue on multimodal interfaces of the Signal Processing journal 86(12):3596–3624

Oh JH, Lee KS, Chang DS, Seo CW, Choi KS (2001) Trec-10 experiments at kaist: Batch filtering and question answering. In: TREC

Reithinger N, Bergweiler S, Engel R, Herzog G, Pfleger N, Romanelli M, Sonntag D (2005) A look under the hood: design and development of the first smartweb system demonstrator. In: ICMI '05: Proceedings of the 7th international conference on Multimodal interfaces, ACM Press, New York, NY, USA, pp 159–166, DOI http://doi.acm.org/10.1145/1088463.1088492

van Schooten B, op den Akker R (2005) Follow-up utterances in QA dialogue. Traitement Automatique des Langues 46(3):181–206

van Schooten B, op den Akker R (2007) Multimodal follow-up questions to multimodal answers in a QA system. In: Tenth international symposium on social communication, Universidad de Oriente Santiago de Cuba, pp 469–474

van Schooten B, Rosset S, Galibert O, Max A, op den Akker R, Illouz G (2007) Handling speech input in the Ritel QA dialogue system. In: Interspeech 2007

van Schooten B, op den Akker R, Rosset S, Galibert O, Max A, Illouz G (2009) Follow-up question handling in the IMIX and Ritel systems: a comparative study. JNLE 15(1):97–118

Small S, Liu T, Shimizu N, Strzalkowski T (2003) HITIQA: an interactive question answering system: A preliminary report. In: Proceedings of the ACL 2003 Workshop on Multilingual Summarization and Question Answering

Theune M, Krahmer E, van Schooten B, op den Akker R, van Hooijdonk C, Marsi E, Bosma W, Hofs D, Nijholt A (2007) Questions, pictures, answers: Introducing pictures in question-answering systems. In: Tenth international symposium on social communication, Universidad de Oriente Santiago de Cuba, pp 450–463

Voorhees EM (2001) Overview of TREC 2001. In: TREC

Voorhees EM (2005) Overview of the TREC 2005 question answering track. Tech. rep., NIST

Wang D, Zhang J, Dai G (2006) A multimodal fusion framework for children's storytelling systems. In: Edutainment, pp 585–588

Willems DJM, Rossignol SYP, Vuurpijl LG (2005) Features for mode detection in natural online pen input. In: BIGS 2005: Proceedings of the 12th Biennial Conference of the International Graphonomics Society, pp 113–117

Witten IH, Frank E (2005) Data Mining: Practical machine learning tools and techniques, 2nd Edition. Morgan Kaufmann

Yang F, Feng J, Di Fabbrizio G (2006) A data driven approach to relevancy recognition for contextual question answering. In: Workshop on Interactive Question Answering, HLT-NAACL 06, pp 33–40

Multidimensional Dialogue Management

Simon Keizer, Harry Bunt and Volha Petukhova

Abstract In natural language dialogue, contributions from the participants are generally multifunctional, that is, in a single contribution a speaker addresses multiple aspects of communication simultaneously. A speaker can directly address the underlying task or activity, such as asking a question about the domain, while at the same time giving feedback about his understanding of the dialogue partner's previous utterance. This phenomenon can be captured elegantly in terms of dialogue acts that are organised in a multidimensional taxonomy. Most state-of-the-art dialogue systems contain a dialogue manager, which takes care of deciding which action to take next in the dialogue, given some form of information state or context model that is monitored and updated during the dialogue. In this chapter a new approach to dialogue management is described, in which the multidimensional nature of communication is supported. A dialogue manager that generates dialogue acts from several dimensions simultaneously allows for a less rigid system behaviour and for more natural interactions with users. As a showcase, the implementation of such a dialogue manager is described, as well as its embedding in the IMIX interactive question answering system.

1 Introduction

Participation in a dialogue requires giving attention to a variety of aspects of conversational interaction. Participants may be interested primarily in pursuing a particular task or activity that motivates their participation in the dialogue, but in

Simon Keizer
Department of Engineering, University of Cambridge, Cambridge, United Kingdom, e-mail: `sk561@cam.ac.uk`

Harry Bunt · Volha Petukhova
Tilburg center for Cognition and Communication, Tilburg University, Tilburg, The Netherlands, e-mail: `{v.petukhova, harry.bunt}@uvt.nl`

order to make the interaction effective, they must make sure that they are understood by the other participants and that also that they themselves understand the others (giving and eliciting feedback); they must coordinate with the other participants regarding who is speaking when (turn management); make their contributions at a certain pace (time management); and should take social conventions into account (e.g. greeting, apologising and thanking when appropriate). Empirical studies have shown that dialogue participants very often address several of these, and other aspects of communication simultaneously, using utterances that are *multifunctional*. Consider, for example, the following dialogue interchange between participants S (system) and U (user):

(1) *U1* : *what is RSI?*
S1 : *RSI stands for Repetitive Strain Injury*
U2 : *but what is it?*
S2 : *Repetitive Strain Injury is an infliction where ...*

When contributing utterance U2, U not only asks a question about something in the task domain (medical information, in the case of IMIX – see Section 4.3), but also indicates that he understood the answer given by S, and that he did not accept S1 as an answer to U1[1]. From a dialogue management perspective, it is important to recognise all three functions in U2 in order to answer U's question in a satisfactory way. Note in particular that merely resolving the anaphora in U2 to 'RSI' from U1 and answering the resulting question would result in a repetition of the answer S1, which would not be helpful.

Computational approaches to dialogue modelling tend to view the multifunctionality of dialogue utterances as a problem, and to not take it into account. Simpler, one-dimensional models may be acceptable for certain very simple dialogue-mediated tasks, but effective and efficient dialogue management should in general exploit, rather than ignore the multifunctionality of natural dialogue utterances.

Most state-of-the-art dialogue managers operate by means of an information state that is updated on the basis of the utterances contributed by the participants, usually represented semantically in terms of dialogue acts of some description. The information state typically contains information about the dialogue history and the progress of the underlying task or activity. Based on the current information state, the dialogue manager decides which dialogue acts to generate for continuing the dialogue. In this chapter, a novel approach to dialogue management is presented which can be categorised as an information state update approach (Traum and Larsson, 2003), but which gives a fundamental account of the multifunctionality of dialogue contributions, both in interpreting user utterances and in generating system utterances.

The remainder of the chapter is organised as follows. First, the phenomenon of multifunctionality in dialogue will be discussed in more detail. Section 2 introduces the semantic framework of Dynamic Interpretation Theory (DIT), including the multidimensional DIT^{++} taxonomy of communicative functions. According to

[1] The two additional functions are in this case due to the multidimensional semantics of the discourse marker 'but' - see (Petukhova and Bunt, 2009a).

this framework, utterances are modelled in terms of dialogue acts that operate on the information states of the participants. Dialogue acts are categorised in a taxonomy that consists of multiple dimensions, each of which represents an aspect of communication that can be addressed independently. The taxonomy thus allows for describing multifunctional dialogue contributions in terms of multiple dialogue acts. Section 3 elaborates on this with a discussion of different forms of multifunctionality that can be captured by this framework, and discusses the relations between communicative functions within and across different kinds of dialogue units.

Section 4 presents the design of a dialogue manager which employs this taxonomy, introducing a context model that represents the information state of the system, as well as a mechanism for the generation of dialogue act combinations based on this context model. These combinations are based on the inputs from several dialogue act agents, software agents that each generate dialogue act candidates for a particular dimension. An additional evaluation agent integrates these candidates into combinations to be presented to the user in the form of multifunctional dialogue contributions. The context update model, which makes use of the central notion of dialogue act preconditions, is discussed in Section 5. The focus then shifts to the dialogue act generation process in Section 6, which discusses logical, strategic, and pragmatic constraints that govern the combination and evaluation of dialogue acts for generating sensibly multifunctional dialogue contributions. The chapter is concluded in Section 7.

2 Semantic and Pragmatic Framework: DIT

2.1 Dimensions and Communicative Functions

In the semantic and pragmatic framework of Dynamic Interpretation Theory (DIT) (Bunt, 1989, 1995, 2000, 2009a), dialogue utterances are interpreted as combinations of dialogue acts, which operate on the information states of the dialogue participants. Dialogue acts have two main components: a *semantic content*, which is to be inserted into, to be extracted from, or to be checked against the current information state; and a *communicative function*, which specifies how an addressee updates his information state with the semantic content when he understands the corresponding aspect of the meaning of a dialogue utterance[2].

DIT takes a multidimensional view on dialogue in the sense that communicative behaviour in dialogue is viewed as the performance of several activities in parallel, such as pursuing an underlying task, providing and eliciting feedback, taking turns, opening and closing topics, and editing one's own or another participant's

[2] Other components of a dialogue act, which are not considered in this chapter, are functional and rhetorical relations between dialogue acts, and feedback relations between dialogue acts and previous utterances (see Bunt et al, 2010).

utterances. Each of these types of communicative activity, performed by dialogue acts, is called a *dimension*. Dialogue acts pertaining to different dimensions are concerned with different types of information; for example, a turn management act is concerned with the allocation of the speaker role; feedback acts are concerned with the processing of each other's utterances, and task-related dialogue acts are concerned with information regarding the task or activity that motivates the dialogue. Each dimension thus corresponds with a particular type of semantic content. A set of dimensions is defined in DIT as follows:

(2) 1. Each dimension is a type of communicative activity which dialogue participants perform by means of dialogue acts with a particular type of semantic content;
2. In each dimension the activities can be performed independently of those in other dimensions.

The second part of (2) requires dimensions to be independent or 'orthogonal', which means that the communicative functions which a segment can have in one dimension are in general not determined by its functions in other dimensions. Petukhova and Bunt (2009a,b) performed a corpus-based investigation which proves the orthogonality of the 10 DIT dimensions, listed below in (3). They also tested these dimensions against 18 existing dialogue act annotation schemes[3], and showed that the distinctions made in these schemas lend empirical support to the use of these dimensions.

(3) 1. **Task(/Activity)**: acts that advance the underlying task or activity;
2. **Auto-Feedback**: acts dealing with the speaker's processing of the addressee's utterances; includes positive and negative feedback acts on various levels of processing (see below);
3. **Allo-Feedback**: acts dealing with the addressee's processing of the speaker's previous utterances (according to the speaker); includes positive and negative feedback-giving acts and feedback elicitation acts on various levels of processing (see below);
4. **Turn Management**: Turn Accept, Turn Grab, Turn Take; Turn Keep, Turn Assign, Turn Release;
5. **Time Management**: Stalling, Pausing;
6. **Partner Processing Management**: Completion, Correct-misspeaking;
7. **Own Processing Management**: Error Signalling, Retraction, Self-correction;
8. **Contact Management**: Contact Check, Contact Indication;
9. **Discourse Structuring**: Opening and Pre-closing, Topic Introduction, Topic Shift, Topic Shift Announcement;
10. **Social Obligations Management**: initiating and responsive dialogue acts for salutation, introducing oneself, thanking, apologising, and valediction.

Concerning the first part of definition (2), Petukhova and Bunt (2009a) examined the types of semantic content addressed by dialogue acts occurring in three different kinds of corpora: the DIAMOND corpus[4], which consists of two-party human-human task-oriented instructional spoken dialogues; the AMI meeting recordings

[3] The schemes DAMSL, SWBD-DAMSL, LIRICS, DIT^{++}, MRDA, Coconut, Verbmobil, HCRC MapTask, Linlin, TRAINS, AMI, SLSA, Alparon, C-Star, Primula, Matis, Chiba and SPAAC.

[4] For more information see Geertzen, Girard, and Morante (2004), The DIAMOND project. Poster at the 8th Workshop on the Semantics and Pragmatics of Dialogue (CATALOG 2004).

corpus[5], which consists of multimodal task-oriented human-human multi-party dialogues; and the OVIS corpus[6], which consists of task-oriented human-computer telephone dialogues. Table 1 presents the distribution of dialogue act tags across the ten dimensions in these corpora.

Table 1 Distribution of dialogue act tags across dimensions for analysed dialogue corpora in (%).

	AMI	DIAMOND	OVIS
Task	31.8	45.1	48.8
Auto-Feedback	20.5	19.1	24.1
Allo-Feedback	0.7	3.8	39.2
Turn Management	50.2	19.9	19.3
Social Obligation Management	0.5	7.8	3.8
Discourse Structuring	2.8	2.3	3.3
Own Communication Management	10.3	0.7	3.4
Time Management	26.7	16.1	10.8
Partner Communication Management	0.3	0.3	0.5
Contact Management	0.1	2.8	12.3

Dimensions classify dialogue acts. What is usually called a 'dialogue act taxonomy' is in fact a taxonomy of the *communicative functions* of dialogue acts, and dialogue act classification is most often understood as the assignment of communicative function tags to segments of dialogue.

Besides classifying dialogue acts, dimensions in DIT also classify those communicative functions which are concerned with one particular type of information, such as a turn grabbing function, which is concerned with the allocation of the speaker role, or an auto-feedback function, which is concerned with the understanding of a previous utterance. Being specific for a particular dimension, these functions are called *dimension-specific*. Most of the dimension-specific functions of DIT are mentioned in (3)[7].

Other communicative functions are not specifically related to any dimension, e.g. it is possible to ask for or provide an answer to a question about any type of semantic content, or request the performance of any type of action. Because they can be used to address any dimension, these communicative functions are called *general-purpose* functions. These functions are divided into *information-transfer* functions, in turn subdivided into information-seeking and information-providing

[5] **A**ugmented **M**ulti-party **I**nteraction (`http://www.amiproject.org/`).

[6] **O**penbaar **V**ervoer **I**nformatie **S**ystem (Public Transport Information System) `http://www.let.rug.nl/~vannoord/Ovis/`

[7] See (Bunt, 2009a) or `http://dit.uvt.nl` for more details.

functions, and *action-discussion* functions as, in turn subdivided into commissive and directive functions, as shown in (4).

(4) • *Information-transfer functions*
  – information-seeking functions: Yes/No-question, WH-question, Alternatives-question, Indirect Yes/No-question, Indirect WH-question, Indirect Alternatives-question, Check, Positive Check, Negative Check;
  – information-providing functions: Inform, Uncertain Inform, Yes/No-answer, WH-answer, Uncertain Yes/No-answer, Uncertain WH-answer, Confirm, Uncertain Confirm, Disconfirm, Uncertain Disconfirm, Disagreement, Correction;
• *Action-discussion functions*
  – commissive functions: Promise, Offer, Accept Request, Decline Request, Accept Suggestion, Decline Suggestion;
  – directive functions: Request, Indirect Request, Instruct, Suggestion, Accept Offer, Decline Offer

The distinction of 10 dimensions, of sets of dimension-specific communicative functions in each of these, of a structured set of general-purpose communicative functions that can be used in any dimension, and the association of an update semantics using structured, 'multidimensional' context models, are the main features that distinguish DIT from other dialogue act based approaches to dialogue modelling, such as PTT (Poesio and Traum, 1997, 1998) and QUD (Ginzburg, 1997; see also Larsson, 1998), or the DAMSL annotation framework (Allen and Core, 1997). Like DIT, PTT analyses dialogue utterances as combinations of dialogue acts, each of which is semantically defined in terms of information state updates. PTT's inventory of dialogue act types is limited to a set of 'core dialogue acts', corresponding to a subset of DIT's general-purpose communicative functions, plus a small set of 'grounding acts', corresponding to a subset of DIT's auto-feedback functions. QUD is focused on the semantics and pragmatics of questions and responses to questions, using a stack of 'questions under discussion' for dealing with multiple questions, as may occur for instance when a question is responded to by a question for clarification. DAMSL considers a wider range of dialogue act types than PTT and QUD, and has been a source of inspiration in designing the DIT^{++} taxonomy of communicative functions, but in the absence of a well-defined notion of dimension and of considerations of implication relations between dialogue acts, it does not offer a satisfactory approach for dealing with the multifunctionality of dialogue segments.

The use of dialogue acts in the interpretation (and generation) of dialogue requires a way to identify stretches of dialogue that correspond to dialogue acts. Dialogues can be decomposed into *turns*, defined as stretches of speech produced by one speaker, bounded by periods of silence of that speaker (Allwood, 2000). Turns consist of one or more *utterances*, linguistically defined stretches of communicative behaviour that have a communicative function. The stretches of behaviour that are relevant for interpretation as dialogue acts often coincide with utterances in this sense, but they may be discontinuous, may overlap, and may even contain parts of more than one turn (see Bunt, 2009c). They therefore do not always correspond to utterances, which is why the notion of a *functional segment* as a minimal stretch of communicative behaviour has been introduced that has a communicative function (and possibly more than one).

As seen above, different aspects of communication that speakers may address simultaneously in their dialogue behaviour are identified and reflected in the DIT taxonomy by virtue of its 10 dimensions. In each functional segment several dialogue acts can be performed, each belonging to a different dimension. A good understanding of the nature of the relations among the multiple functions that a dialogue unit may have, and how these units relate to other units in dialogue, opens the way for defining a computational model for the interpretation and generation of multifunctional dialogue utterances. In the next section the forms of multifunctionality are discussed that occur in natural dialogue and the relations between the communicative functions of a multifunctional segment.

3 Multifunctionality

3.1 Relations Between Communicative Functions

Bunt (2010a) defines two types of implication relations that may exist between dialogue acts and between communicative functions: *entailment* and *implicature*. These relations are introduced here, plus the related notion of *mutual exclusion*, and how these relations turn up especially in relation to feedback and indirectness is considered.

Entailment relations between dialogue acts and between communicative functions are defined as follows, where $||A||_M$ is used to denote the interpretation of dialogue act A given the context model M or, equivalently, the update effects on context model M caused by dialogue act A:

(5) a. A dialogue act A_1 entails a dialogue act A_2 if for any context model M, the update effects $||A_1||_M$ on M that would be caused by A_1 have the update effects $||A_2||_M$ that would be caused by A_2 as logical consequences.

b. A communicative function F_1 entails a communicative function F_2 iff a dialogue act with communicative function F_1 entails the dialogue act with communicative function F_2 and the same semantic content.

One type of entailment relation is the one between dialogue acts within the same dimension that differ in specificity, more specific dialogue acts entailing less specific ones. For example, Agreement and Disagreement entail Inform, and Confirm and Disconfirm entail Yes/No-Answer. This type of intra-dimension entailment relation has been called *functional subsumption* (Bunt, 2009b).

Since dialogue acts in different dimensions are concerned with different types of information, no entailment relations are likely to exist between them. An exception is the relation between responsive non-feedback acts on the one hand and auto- and allo-feedback acts on the other. This type of entailment occurs because a dialogue act which responds to a dialogue act from another participant, such as accepting an offer, answering a question, or accepting an apology, presuppose the successful processing of the utterance which expressed that dialogue act that is responded to.

Implicature relations between dialogue acts and between communicative functions are defined as follows:

(6) a. A dialogue act A_1 *implicates* a dialogue act A_2 if for any context model M, the update effects $||A_1||_M$ of A_1 on M have the update effects $||A_2||_M$ of A_2 as conversational implicatures.
b. A communicative function F_1 implicates a communicative function F_2 if a dialogue act with communicative function F_1 implicates the dialogue act with communicative function F_2 and the same semantic content.

Implicated functions are not expressed explicitly, through the features of expressions, but can be inferred as being likely in a given context. Implicated functions are intended to be recognised, and like all conversational implicatures, can be cancelled. Examples of implicated functions are:

1. an expression of thanks implicating positive feedback (at all levels of processing) concerning the previous utterance(s) of the addressee;
2. positive feedback implied by moving on to a new, relevant topic; more generally, by any relevant continuation of the dialogue;
3. negative feedback, implied by shifting to an *un*related topic; more generally, by any *ir*relevant continuation of the dialogue.

Two important cases of implicated multifunctionality are formed by indirect dialogue acts and by feedback acts addressing different levels of processing.

Indirect dialogue acts occur when a speaker uses a linguistic form that literally expresses one type of dialogue act, but in context means something else. Questions of the form *Do you know [X]* illustrate this: while an utterance of this form would literally seem to ask whether the addressee possess the information $[X]$, it is most often used to request to provide the information $[X]$, if possible. (*If you know* $[X]$, *please tell me.*) This makes such a question a *conditional request*.

The DIT taxonomy views indirect dialogue acts as having a communicative function which is slightly different from that of the corresponding direct form, because their performance is thought to have slightly different effects on information states. For example, the difference between *Where is Lee's office?* (WH-Question) and *Do you know where Lee's office is?* (Indirect WH-Question) would be that in the indirect version the speaker does not express an assumption that the addressee knows the answer to his question, whereas in the direct version he does[8].

---

[8] This approach is taken in release 4 of the DIT taxonomy, which is the one that was used at the time of the IMIX project (see `http://dit.uvt.nl/dit4`). A slightly different approach is taken

Relations among feedback levels. In DIT, five levels of processing are distinguished at which feedback acts may apply. Ordered from low to high, these are:

(7) *attention* $<$ *perception* $<$ *interpretation* $<$ *evaluation* $<$ *execution*

The ordering of these levels of processing gives rise to entailment relations between feedback acts at different levels, both in the Auto- and in the Allo-Feedback dimension. In the case of positive feedback, an act at level L_i entails positive feedback at all levels L_j where $i > j$; for negative feedback the entailment relation works in the opposite direction.

The existence of levels of feedback also gives rise to conversational implicatures. The Cooperation Principle (Grice, 1975), in particular the Maxim of Quantity, requires a speaker who provides feedback about his processing of a previous utterance to do this in an optimally informative way; for positive feedback this is at the highest level of successful processing; for negative feedback at the lowest level of unsuccessful processing. Therefore, positive feedback at level L_i implicates negative feedback at all levels L_j where $i < j$; negative feedback at level L_i implicates positive feedback at all levels L_j where $j < i$; for instance, a negative feedback act signalling a semantic interpretation problem implicates positive feedback concerning perception.

Mutual exclusion. A sort of opposite entailment is the relation of *mutual exclusion*:

(8) Two dialogue acts A_1 and A_2 mutually exclude each other if the application of both the updates that would be caused by A_1 and by A_2 would result in an inconsistent state of the context model, i.e. a state in which some proposition P is true as well as its negation.

Assuming dialogue participants to be rational agents, with consistent context models[9], a functional segment of their communicative behaviour cannot have two mutually exclusive functions.

The DIT tagset has been designed in such a way that two communicative functions which can be applied in the same dimension either (1) mutually exclude each other, or (2) one entails the other. Consider, for example, the Time Management dimension. The speaker may need a little time to formulate his utterance, for instance because while speaking he is looking at a computer screen trying to identify certain information that he was asked about; in this situation he may well perform a Stalling act (... *let me see...*, *ehm...*) in order to indicate this to the addressee. In another situation he may want to suspend the dialogue for a while, for instance because an urgent phone call comes in; in this case he may well perform a Pausing act to signal this (like *Just a minute*). Stalling and pausing acts are mutually exclusive: they cannot both apply to one and the same segment, since a speaker either suspends the dialogue at that point or he continues.

---

in release 5 (see `http://dit.uvt.nl`), which makes uses of so-called *function qualifiers*, such as *conditional*, which can be used in this example to characterize the communicative function as a conditional request; for more details see (Petukhova and Bunt, 2010).

[9] The terms 'context model' and 'information state' are used interchangeably in this chapter.

The next subsection shows how the relations between communicative functions influence the possible forms of multifunctionality in functional segments and larger units in dialogue.

3.2 Types of Multifunctionality in Dialogue Units

A functional segment may have more than one communicative function for the following reasons:

1. its surface characteristics, such as wording, prosody and accompanying non-verbal signals encode more than one function;
2. one of its functions entails or implicates another one.

In the first case we speak of *independent* multifunctionality; in the second case of *entailed* or *implicated* multifunctionality, respectively. Note that, since any two communicative functions in a DIT dimension either entail or mutually exclude each other, independent multifunctionality can only occur if the dialogue acts involved belong to different dimensions.

To examine the forms of multifunctionality that occur in natural dialogue a corpus analysis was performed, using human-human multi-party interactions (AMI meetings). Three scenario-based meetings were selected containing 17,335 words. Dialogue contributions were segmented at turn level (776 turns), at utterance level (2,620 utterances), and at the finer level of functional segments (3,897 functional segments). The data was annotated according to the DIT++ annotation schema[10].

3.2.1 Multifunctionality in Segments

The analysis shows that functional segments often display independent multifunctionality in the form of encoding several functions in different dimensions. For example:

(9) *B1: any of you anything to add to that at all?*
A1: no
D1: i'll add it later in my presentation

In utterance B1, which is accompanied by the speaker looking around, a question is expressed (by word order and prosody) as well as a Turn Release act (expressed by *'any of you'* in combination with looking around).

Utterance A1 in (9) illustrates the possible entailed multifunctionality of a segment: the speaker answers question B1, and by implication provides positive feedback concerning his understanding of that utterance.

[10] See Bunt (2009a), or `http://dit.uvt.nl/`.

Table 2 gives an overview of the co-occurrences of communicative functions across dimensions for a functional segment, both with and without taking entailed and implicated functions into account.[11] The zero figures on the diagonal confirm that functions which address the same dimension co-occur only when there is an entailment or implicature relation between them, as happens all the time for Auto- and Allo-Feedback functions.

Table 2 Co-occurrences of communicative functions across dimensions in the AMI corpus, expressed in relative frequency in %, implicated and entailed functions excluded and included. (To be read as follows: percentage of segments having a communicative functions in the dimension corresponding to the column, which also has a function in the dimension corresponding to the row).

	Form	Task	Auto-F.	Allo-F.	Turn M.	Time	DS	Contact	OCM	PCM	SOM
Task	Indep.	0.0	1.1	0.0	2.2	0.1	19.6	0.0	3.8	0.0	0.0
	Implied	49.8	47.9	24.9	97.5	2.4	31.5	0.4	69.6	0.1	0.7
Auto-F.	Indep.	0.7	0.0	0.0	11.0	0.6	1.9	11.1	0.8	0.0	0.0
	Implied	38.9	0.0	0.0	88.7	11.4	11.2	20.2	11.7	65.0	8.7
Allo-F.	Indep.	0.0	0.0	0.0	0.1	0.0	0.0	0.0	0.0	0.0	0.0
	Implied	24.9	0.0	0.0	94.8	35.7	2.1	1.2	7.9	0.7	0.3
Turn M.	Indep.	3.4	26.9	6.7	0.0	28.6	12.4	7.4	4.8	18.2	6.7
	Implied	76.0	66.2	19.4	0.0	42.9	14.6	13.8	99.6	27.3	10.5
Time M.	Indep.	0.1	0.7	0.0	44.9	0.0	4.7	0.0	1.3	0.0	0.0
	Implied	28.2	11.3	7.8	98.6	0.0	1.7	0.0	83.2	0.5	0.0
DS	Indep.	0.1	0.4	0.0	0.3	0.0	0.0	0.9	0.0	0.0	6.7
	Implied	3.2	58.3	29.1	87.5	4.9	4.6	25.0	3.7	0.0	12.5
Contact M.	Indep.	1.7	0.3	0.0	3.6	0.5	3.7	0.0	0.0	0.0	1.3
	Implied	2.4	97.1	1.6	98.8	0.5	2.4	0.0	0.3	0.0	3.7
OCM	Indep.	1.2	0.4	0.0	2.8	0.5	0.0	0.0	0.0	0.0	6.7
	Implied	82.2	2.8	2.5	96.9	7.8	3.9	13.5	0.0	0.9	7.6
PCM	Indep.	0.0	0.0	0.0	0.3	0.0	0.0	0.0	0.0	0.0	0.0
	Implied	11.8	65.0	11.8	79.1	12.2	0.0	0.0	0.0	0.0	0.0
SOM	Indep.t	0.0	0.0	0.0	0.2	0.0	0.0	2.7	0.3	0.0	0.0
	Implied	0.7	80.0	10.0	90.0	0.0	30.0	3.9	2.0	0.0	0.0

Some combinations of functions are relatively frequent, e.g. time- and turn management acts often co-occur. A speaker who wants to win some time to gather his thoughts and wants to continue in the sender role, may signal both by slowing down and using fillers, expressing both a Stalling and a Turn Keeping act. (Stalling behaviour does not *always* have a turn-keeping function, however; an extensive amount of stallings accompanied by relatively long pauses may for instance be intended to elicit support for completing an utterance.)

[11] Only cross-dimension entailments are taken into consideration, no within-dimension functional subsumption.

Table 2 shows that the frequency of co-occurrence of functions from any two dimensions D1 and D2 increases spectacularly when entailed and implicated functions are taken into account; for example, 2.2% of the dialogue acts in the Task dimension have an *independent* Turn Management function, while 95.3% of them have an *implicated* Turn Management function. Questions, for instance, which often belong to the Task dimension, much of the time have an implicated Turn Release or Turn Assign function, allowing the question to be answered. (This implicature may however be cancelled or suspended when the speaker does not stop talking right away after asking the question.) This supports the claim in Bunt (2009b) that entailment and implicature relations are the main source of multifunctionality in functional segments, more so than independent multifunctionality.

3.2.2 Multifunctionality in Segment Sequences

Dialogue participants of course do not limit their dialogue contributions to functional segments; much of the time they produce utterances containing several segments. Functional segments may overlap, either in the form of one segment interrupting another, as in (10a), or as one segment being part of another, as in (10b).

(10) a. *Twenty five Euros for a remote...* ***how much is that locally in pounds?*** *is too much to buy a new one*
b. *D1: Which is the clunky one on the left or on the right?*
C1: ***The clunky one*** *is the one on the right*

The part marked in bold in utterance C1 in (10b) forms explicit positive feedback on understanding and accepting the preceding question. The shorter form *'on the right'* would instead have provided positive feedback implicitly, by entailment.

It is often possible to add explicitly what would otherwise be entailed or implicated without being redundant. For example, positive feedback implicated by shifting to a relevant new topic may be expressed explicitly by means of discourse markers, such as *'and then', 'okay then', 'next'* (see Petukhova and Bunt, 2009a). For example:

(11) *D1: This idea focuses on <u>the twenty five age group.</u>*
B1: Are we aiming at ***<u>a fairly young market</u> then****?*

Functional segments following each other within a turn give rise to multifunctionality at turn level. Segment sequences of length 2 were analysed for the most frequently occurring patterns of communicative function combinations; see Table 3. The results are markedly different from those for single segments (Table 2); an obvious, expected difference is that subsequent segments may express different acts in the same dimension.

The co-occurrence scores for Turn Management, Task, and Auto-Feedback with other dimensions are relatively high. Task-related functional segments are frequently preceded or followed by Turn Management or Auto-Feedback segments

Table 3 Co-occurrences of communicative functions across dimensions in a sequence of two functional segments in one turn, expressed in relative frequency in %.

within	Task	Auto-F.	Allo-F.	Turn M.	Time M.	DS	Contact M.	OCM	PCM	SOM
Task	26.5	36.5	33.3	33.5	42.4	0.0	15.4	21.6	20.0	46.7
Auto-F.	15.9	24.8	9.9	16.7	17.2	33.3	19.2	8.0	30.0	13.3
Allo-F.	0.4	1.1	6.6	0.6	0.6	0.0	0.0	0.5	0.0	0.0
TurnM.	59.7	38.1	36.7	53.0	44.2	15.3	61.5	69.9	50.0	33.3
TimeM.	27.9	20.4	20.0	30.9	18.8	0.0	15.4	55.4	0.0	26.7
ContactM.	0.0	0.1	0.0	0.1	0.0	34.2	0.0	0.0	0.0	54.6
DS	0.49	1.2	0.0	0.6	0.6	15.0	7.6	0.5	0.0	0.0
OCM	9.9	8.0	6.7	11.3	13.9	0.0	7.7	9.5	0.0	0.0
PCM	0.4	0.4	0.0	0.1	0.1	0.0	0.0	0.3	0.0	0.0
SOM	0.2	0.6	0.0	0.3	0.1	33.3	0.0	0.5	0.0	6.7

(or segments that have functions in both these dimensions). A frequent pattern for constructing a turn is first performing a turn-initial act (Turn Take, Accept or Grab) combined with, or followed by an Auto-Feedback act, and one or more segments in another dimension, and closing with a turn-final act. This pattern occurs in 49.9% of all turns. For example:

(12) *B1: well (Turn Take + Negative Auto-Feedback Evaluation)*
B2: twenty five euro is about eighteen pounds, isn't it? (Task-related Check Question)
D1: um (Turn Take+Stalling)
D2: yep (Task-related Confirm + Turn Release)

The next section will consider how these findings contribute to the design of a dialogue manager that aims to generate coherent and natural deliberately multifunctional dialogue contributions.

4 Design of a Multidimensional Dialogue Manager

Having outlined the semantic framework of DIT, the overall design can be introduced of a dialogue manager in which the multidimensional nature of communication is supported, both in interpreting user actions as well as generating system actions. Although the interpretation module is an essential component of a dialogue system, the focus here will be on the generation of dialogue acts.

4.1 Context Model

Central to the dialogue manager is the context model, representing the information state of the dialogue system and that of the user as seen by the system. This context model is closely related to the dialogue act taxonomy in the sense that dialogue acts are defined as context update operators, and have preconditions specified in terms of properties of the context model of the speaker. It can be argued that the context model in fact serves as a formal semantics for dialogue annotation, such an annotation being a kind of underspecified semantic representation (Bunt and Keizer, 2005; Bunt, 2009c).

Given that dialogue acts are categorised in dimensions, each of which indicate a different type of information addressed by an act, these dimensions are also partly reflected in the structure of the context model. In general, the context model should contain all information that is relevant for the interpretation and generation of dialogue acts (see Bunt, 2010a for details). These and other considerations related to convenience of representation have lead to the following general structure:

1. *Linguistic Context*: information about the utterances produced in the dialogue so far (a kind of 'extended dialogue history'); information about planned system dialogue acts (the 'dialogue future');
2. *Semantic Context*: information about the underlying task or activity, including assumptions about the dialogue partner's information;
3. *Cognitive Context*: the current processing states of both participants (on the levels of attention, perception, understanding, evaluation, and execution);
4. *Physical and Perceptual Context*: the perceptible aspects of the communication process and the task/domain;
5. *Social Context*: current communicative pressures.

During a dialogue, the context model is updated on the basis of the dialogue contributions of the participants. On the other hand, the updated context model gives rise to the production of dialogue contributions, in particular dialogue system responses.

4.2 Dialogue Act Agents

The multidimensional view on dialogue modelling in DIT suggests that in generating dialogue behaviour, participants select dialogue acts from different dimensions simultaneously and independently, and combine them into multifunctional utterances. The dialogue manager presented here is therefore designed to consist of multiple *dialogue act agents* that operate in parallel on the context model. Each of these agents is dedicated to generating dialogue acts from one particular dimension. An additional *evaluation agent* combines the dialogue act candidates generated by the dialogue act agents into system contributions. The orthogonality of the dimensions ensures that the dialogue act agents can operate independently. As seen in Section 3.1 however, there are relationships between communicative functions, and these impose constraints on the process of combining dialogue acts. The design of the evaluation agent, therefore, is not a trivial issue.

The architecture of the dialogue manager is given in Figure 1, where four components of the context model and five dialogue act agents are shown. The results from speech and language understanding, in particular Dialogue Act Recognition (DAR) are recorded in the Linguistic Context. The Context Manager carries out the necessary updates and reports any resulting inconsistencies. The combinations of dialogue acts resulting from the dialogue act generation and evaluation process are passed on to the Natural Language Generation (NLG) component.

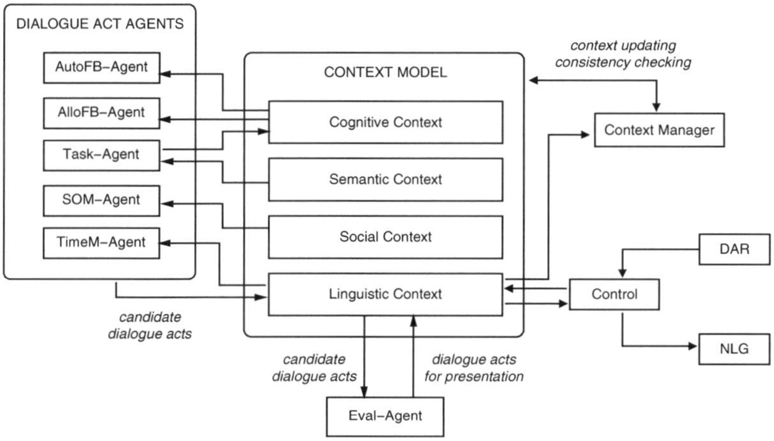

Fig. 1 Multidimensional dialogue manager architecture.

All the dialogue act agents continuously monitor the context model and, if appropriate, try to generate candidate dialogue acts from their associated dimension. This process of monitoring and act generation is modelled through a triggering mechanism: if the information state satisfies the agent's *triggering conditions*, i.e., if there is a motivation for generating a dialogue act from that particular dimension, the corresponding agent is triggered and tries to generate such a dialogue act. For example, the Auto-Feedback Agent is triggered if a processing problem is recorded in the Own Processing State of the Cognitive Context. The agent then tries to generate a negative auto-feedback act in order to solve the processing problem (e.g., *'Could you repeat that please?'* or *'Did you say "five"?'*). The Auto-Feedback Agent may also be triggered if it has reason to believe that the user is not sure if the system understood a previous utterance, or simply if it has not given any explicit positive feedback for some time. In these cases of triggering, the agent tries to generate a positive auto-feedback act.

Hence the dialogue act generation process involves 10 dialogue act agents that operate in parallel on the context model, and an evaluation agent. The dialogue acts generated by these agents are kept in the linguistic context as candidates (hence the 'dialogue future'). The selection of dialogue acts from different dimensions may happen independently, but for their order of performance and their combination, the relative importance of the dimensions at the given point in the dialogue has to be taken into account.

The evaluation agent monitors the list of candidates and decides which of them can be combined into a multifunctional system utterance for generation, and when. Some of the dialogue act candidates may have higher priority and should be generated at once, some may be stored for possible generation in later system turns, and some will already be performed by implication through the performance of other candidate acts. As will be discussed in Section 6, the process of combining and evaluating dialogue acts for generation involves logical, strategic, and pragmatic considerations, and allows for the flexible implementation of different dialogue strategies and styles of communication. A similar argument is mentioned by Stent (2002), describing a dialogue manager that consists of three independent agents operating in parallel. The 'organisation of conversation acts into coherent and natural dialogue contributions' is taken care of by one of these agents, called the generation manager. The distinction between the processes of 'contribution planning' and 'contribution structuring' has some similarity with the distinction between the dialogue act agents (over-)generating dialogue acts and the evaluation agent selecting and combining the resulting candidates. However, contribution structuring deals with interrelationships between the *levels* of conversation acts, whereas the evaluation agent operates on the basis of interdependencies between *dimensions* of dialogue acts.

Another approach to dialogue management in which a multi-agent approach is taken, is the JASPIS speech application architecture for adaptive and flexible human-computer interaction (Turunen et al, 2005). The system uses so-called 'Evaluators' that determine which agents should be selected for different interaction tasks, based on evaluation scores. Part of an Evaluator's task may be to decide on a particular dialogue strategy by selecting a corresponding dialogue agent. This kind of evaluation process however is different from the approach in that it only involves selecting between alternative agents for the same task.

4.3 Application: Dialogue Management for Interactive QA

As part of the IMIX research programme, a proof-of-concept demonstration system was developed, in which components from all partners were integrated. The main goal of this endeavour was to stimulate multidisciplinary collaboration and knowledge transfer between highly autonomous projects, rather than developing an extensively tested and evaluated first prototype. The PARADIME (PARallel Agent-base DIalogue Management Engine) project contributed an initial implementation of the multidimensional dialogue manager discussed in this chapter to this interactive question-answering (QA) demonstration system (Keizer and Bunt, 2006).

The task-domain at hand concerns encyclopedic information in the medical domain, in particular RSI (Repetitive Strain Injury). The system consists of several input analysis modules (ASR, syntactic analysis in terms of dependency trees, and shallow semantic tagging), three different QA modules that take self-contained domain questions and return answers retrieved from several electronic documents containing textual data in the medical domain, and a presentation module that takes the output from the dialogue manager, possibly combining any QA-answers to be presented, into a multimodal system utterance (see also Hofs et al, Chapter 3, this volume).

4.3.1 Functionality

In the design of this proof-of-concept system the role of the dialogue management module is to provide support for interactive and coherent dialogues, in which problems can be solved about both communication and QA processes. In interaction with the user, the system plays the role of an Information Search Assistant (ISA). This HCI metaphor posits that the dialogue system is not an expert on the domain, but merely assists the user in formulating questions about the domain that will lead to answers from the QA modules satisfying the user's information need (Op den Akker et al, 2005).

For this version of the dialogue manager limited system functionality was defined, and following from that a simplified version of the context model and the dialogue act taxonomy. This resulted in a dialogue manager containing four dialogue act agents: a *Task-Oriented (TO) Agent*, an *Auto-Feedback (AUF) Agent*, an *Allo-Feedback (AUF) Agent*, and a *Social Obligations Management (SOM) Agent*. In addition, a very simple evaluation agent takes care of merging candidate dialogue acts for output presentation.

4.3.2 The Task-Oriented Agent

The Task-Oriented (TO) Agent is dedicated to the generation of task-specific dialogue acts, which in practice involves answer acts intended to satisfy the user's information need as indicated through his/her domain questions. The agent is triggered if a new information need is recorded in the Semantic Context. Once it has been triggered, the agent sends a request to the QA modules to come up with answers to a question asked, and evaluates the returned results, based on the number of answers received and their confidence scores. If the QA did not find any answers or if the answers produced had confidence scores that were all below some *lower threshold*, the TO-Agent will not generate a dialogue act, but report an execution problem in the Own Processing State of the Cognitive Context (which causes the Auto-Feedback Agent to be triggered, see Section 4.3.3). Otherwise, the TO-Agent tries to make a selection from the QA answers to be presented to the user. If this selection turns out to contain too many answers, again, an execution problem is written in the Cognitive Context (the question might have been too general to be answerable). Otherwise, the selection is included in an answer dialogue act, either a WH-Answer, or Uncertain WH-Answer in case the confidence scores are below some *upper threshold*. The selection is narrowed down further if there is a subselection of answers with confidences that are significantly higher than those of the other answers in the selection.

4.3.3 The Auto-Feedback-Agent

The Auto-feedback (AUF) Agent is dedicated to the generation of auto-feedback dialogue acts. In this preliminary version it produces negative auto-feedback acts on the levels of interpretation (*'I didn't understand what you said'*), evaluation (*'I do not know what to do with this'*) and execution (*'I could not find any answers to your question'*). It may also decide to occasionally give positive feedback to the user. For future versions, this agent should also be able to generate articulate feedback acts. For example, in S2 of fragment 13, such a feedback act is generated with the purpose of resolving a reference resolution problem:

(13) *U1 :* *what is RSI?*
S1 : *RSI (repetitive strain injury) is a pain or discomfort caused by small repetitive movements or tensions.*
U2 : *how can it be prevented?*
S2 : *do you mean 'RSI' or 'pain'?*

4.3.4 Social Obligations Management Agent

The Social Obligations Management (SOM) Agent is dedicated to the generation of social obligations management acts, some of which might also serve as discourse structuring acts (opening resp. closing through a greeting resp. valediction act). The agent is triggered if communicative pressures are recorded in the Social Context. Currently it only responds to reactive pressures as caused by initiative greetings and goodbyes from the user.

5 Context Specification and Update Mechanisms

In the previous section the context model was introduced as a five-component representation of the information state of the system. In developing the detailed specification that will be described in this section, the main guideline has been the principle that the context model should contain the information that is relevant for interpreting and generating dialogue acts. Since the dialogue acts considered are from the DIT multidimensional taxonomy, the context model reflects the dimensions and allows expression of the preconditions that define the dialogue acts.

5.1 Specification of the Context Model

The feature structure in Figure 2 gives a detailed specification of the context model, following the general structure outlined in the previous section:

- The contents of the **linguistic context** are represented in terms of functional segments, as introduced in Section 2, each of which contains information about the surface realisation as well as the semantic interpretation in terms of dialogue acts.
- The **semantic context** contains the systems's beliefs about the task; the user's beliefs concerning the task (as viewed by the system); and their assumed shared beliefs (or 'common ground' – see Section 5.3) concerning the task.
- The **cognitive context** keeps track of the processing states of the system and the user (according to the system), and specifies the nature of any processing problems, thereby referring to elements in the linguistic context.
- The **perceptual context** contains information regarding the participants' presence and readiness, enabling the system's contact management capabilities.
- The **social context** contains information about current reactive and interactive pressures (Bunt, 1995), in particular in relation to social obligations.

The representation of each of these components consists of three parts: (1) the system's beliefs about the task, about the processing of previous utterances, or about certain aspects of the interactive situation; (2) the user's beliefs of the same kind, according to the system; and (3) the beliefs of the same kind which the system assumes to be shared (or 'grounded') with the user. Note that part (2) introduces full recursion in each component of the context model; it depends on the kind of application whether this is indeed necessary (see Bunt, 2000). The context model implemented in the PARADIME system is limited in this respect.

5.2 Levels of Processing and Feedback

As mentioned above (see (7)), DIT distinguishes five levels of understanding that participants in a dialogue may reach in processing each other's utterances. The lowest level is that of *Attention* giving, and is not considered here. In a dialogue system, the level of *Perception* is associated with successful speech recognition. *Interpretation* corresponds to being able to identify the intentions of the speaker in producing the utterance, i.e., to recognise the dialogue act(s) performed with the utterance. *Evaluation* level understanding is concerned with examining whether the beliefs that result from the interpretation of the input (viz. the preconditions of the recognised dialogue acts) are consistent with the current information state. Reaching the *Execution* level means being able to do something with the result of the evaluation. For example, in the case of a question, it consists of finding the information asked for; in the case of an answer, it means taking over the content of the answer.

$$
\begin{bmatrix}
\mathit{LingContext}: & \begin{bmatrix} \mathit{system}: & \begin{bmatrix} \mathit{dialogue\_history}: \langle \mathit{functional\_segments} \rangle \\ \mathit{current\_segment}: \begin{bmatrix} \mathit{start}: \langle \mathit{timepoint} \rangle \\ \mathit{end}: \langle \mathit{timepoint} \rangle \\ \mathit{expression}: \langle \mathit{verbatim/nonverbal\,form} \rangle \\ \mathit{dial\_acts}: \{ \langle \begin{bmatrix} \mathit{dimension}: \langle \mathit{dim} \rangle \\ \mathit{comm\_function}: \langle \mathit{cf} \rangle \\ \mathit{sem\_content}: \langle \mathit{content} \rangle \end{bmatrix} \rangle \} \\ \mathit{sender}: \mathit{system} | \mathit{user} \end{bmatrix} \\ \mathit{dialogue\_future}: \langle \mathit{discourse\_plan} \rangle \end{bmatrix} \\ \mathit{user}: \langle \mathit{partner's\ linguistic\ context} \rangle & \\ \mathit{shared}: \langle \mathit{shared\ linguistic\ context} \rangle & \end{bmatrix} \\
\mathit{SemContext}: & \begin{bmatrix} \mathit{system}: \begin{bmatrix} \mathit{task\_model}: \langle \mathit{beliefs} \rangle \end{bmatrix} \\ \mathit{user}: \langle \mathit{partner's\ semanttic\ context} \rangle \\ \mathit{shared}: \langle \mathit{believed\ mutual\ beliefs\ about\ dialogue\ task} \rangle \end{bmatrix} \\
\mathit{CogContext}: & \begin{bmatrix} \mathit{system}: & \begin{bmatrix} \mathit{own\_proc\_state}: \begin{bmatrix} \mathit{proc\_level}: \mathit{att} | \mathit{perc} | \mathit{int} | \mathit{eval} | \mathit{exec} \\ \mathit{proc\_problem}: \mathit{yes} | \mathit{no} \\ \mathit{problem\_input}: x \\ \mathit{time\_need}: \mathit{negligible} | \mathit{small} | \mathit{substantial} \end{bmatrix} \\ \mathit{partner\_proc\_state}: \begin{bmatrix} \mathit{proc\_level}: \mathit{att} | \mathit{perc} | \mathit{int} | \mathit{eval} | \mathit{exec} \\ \mathit{proc\_problem}: \mathit{yes} | \mathit{no} \\ \mathit{problem\_input}: y \\ \mathit{time\_need}: \mathit{negligible} | \mathit{small} | \mathit{substantial} \end{bmatrix} \end{bmatrix} \\ \mathit{user}: \langle \mathit{user's\ cognitive\ context} \rangle & \\ \mathit{shared\_CogContext}: \langle \mathit{beliefs\ about\ mutual\ understanding} \rangle & \end{bmatrix} \\
\mathit{PercPhysContext}: & \begin{bmatrix} \mathit{system}: \begin{bmatrix} \mathit{presence}: \mathit{positive} | \mathit{negative} \\ \mathit{readiness}: \mathit{positive} | \mathit{negative} \end{bmatrix} \\ \mathit{user}: \langle \mathit{user's\ perceptual/physical\ context} \rangle \\ \mathit{shared}: \langle \mathit{shared\ perceptual/physical\ context} \rangle \end{bmatrix} \\
\mathit{SocContext}: & \begin{bmatrix} \mathit{system}: \begin{bmatrix} \mathit{reactive\_pressure}: \langle \mathit{dial\_act} \rangle \\ \mathit{interactive\_pressure}: \langle \mathit{dial\_act} \rangle \end{bmatrix} \\ \mathit{user}: \langle \mathit{user's\ social\ context} \rangle \\ \mathit{shared}: \langle \mathit{shared\ social\ context} \rangle \end{bmatrix}
\end{bmatrix}
$$

Fig. 2 Feature structure representation of the context model.

5.3 Grounding

Part of the information in the context model is related to the processes of information transfer and grounding in terms of beliefs and goals. The update mechanisms involved in these processes have been specifically developed for the DIT framework, and include a specific take on grounding (Bunt et al, 2007). Four types of belief are distinguished, represented by means of the operators (*believe, weakly believe, mutually believe,* and *know value of*); moreover, the *want* operator represents a participant's goal. These operators are used to express the preconditions which characterise a communicative function. As a dialogue evolves, new beliefs are created; weak beliefs may become strengthened to firm beliefs; and beliefs and goals may be cancelled as the result of understanding dialogue contributions at various levels of understanding. These changes are implemented by means of the operations of *creation*, *strengthening*, *adoption*, and *cancellation*. Dialogue acts have effects on addressees through their understanding of the speaker's contributions, and have effects on all participants (speaker and addressee, in a two-party dialogue) because of their shared assumption that the speaker expects to be understood and believed, unless there is evidence to the contrary (see Bunt, Morante, and Keizer, 2007).

5.4 Context Update Model

Building on an initial version formulated by Keizer and Morante (2006), the context update algorithm takes as input an abstract representation of the utterance produced in the dialogue, either by the user or by the system. When the system processes a user utterance, the language analysis components produce that abstract representation, generally in terms of dialogue acts, except if the system had problems in identifying the dialogue acts performed in the utterance. The starting point of updating the context model with a dialogue act produced by the user consists of creating beliefs about the dialogue act preconditions, which connect the dialogue act to the context model. For example, the user asking a question leads to the system's beliefs that the user wants to know something and that the user believes that the system knows that something, that something being determined by the semantic content of the dialogue act. When the system produces an utterance, the multi-agent dialogue act generator produces the abstract representation in terms of (sets of) dialogue acts. The basis for the system to generate a dialogue act is formed by the information in the context model, the dialogue act preconditions that have to be satisfied, and further constraints on the overall generation process (see Section 6).

5.4.1 Processing a User Utterance

When a user utterance is processed, changes are made in the context model, depending on the meta-information resulting from the process of recognition and understanding, and on the dialogue act information in case of successful understanding. Dialogue acts in the task/activity dimension (creating beliefs about the user's beliefs and goals related to the task) and negative feedback acts (cancelling such beliefs) will provoke changes in the Semantic Context. Problems encountered by the system in processing the user utterance, and negative allo-feedback acts will provoke changes in the Own Processing State of the Cognitive Context. Negative auto-feedback acts, or reasoning about the information in the context model after any kind of dialogue act, will provoke changes in the Partner Processing State of the Cognitive Context. Other cases are discourse structuring acts which change the information about the conversational structure in the Linguistic Context, and SOM acts that create communicative pressures in the Social Context.

5.4.2 Processing a System Utterance

When the system has produced an utterance, the dialogue acts representing that utterance also cause a context update.

The generation of dialogue acts generally provokes the creation of mutual beliefs about (the system's) expectations of (the user's) understanding and adoption, based on the system's existing beliefs corresponding to the preconditions of the dialogue act (i.e., the beliefs that caused this dialogue act to be generated).

Negative auto-feedback acts deal with processing problems as recorded in the Own Processing State of the Cognitive Context, but actually solving those problems depends on the user's response. Discourse structuring acts change the conversational state, and responsive SOM acts release the communicative pressures that triggered those acts.

6 Constraints on Generating Combinations of Dialogue Acts

As outlined in Subsection 4.3, the dialogue manager presented here is designed to generate multifunctional dialogue contributions through the operation of multiple dialogue act agents and an additional evaluation agent. Each dialogue act agent concentrates on a particular dimension of communication, generating dialogue act candidates for that dimension only, and the evaluation agent produces valid combinations from these candidates to be converted into natural language utterances, possibly with non-verbal components. The design of these agents is driven by two types of constraints on dialogue act combinations. The entailment relations that may exist between dialogue acts give rise to logical constraints, whereas implicatures between acts give rise to pragmatic constraints. Both types of constraints are used to avoid inconsistencies in dialogue act combinations, ensuring rational behaviour, but also to implement different dialogue strategies and styles of communication (Keizer and Bunt, 2007).

The individual dialogue act agents are responsible for making sure that no two dialogue acts from the same dimension can be generated simultaneously, except via an entailment relationship. Detecting and resolving logical conflicts due to cross-dimension entailments on the other hand are the evaluation agent's responsibility. The implementation of pragmatic constraints is for the most part concentrated within the evaluation agent.

6.1 Logical Constraints

From a logical point of view, two communicative functions cannot be applied to one and the same semantic content if they have logical conflicts in their preconditions and/or entailments. An offer cannot be accepted and rejected at the same time, because the speaker cannot at the same time want to do the action and not want to do the action. An answer to a question, entailing positive auto-feedback at interpretation level, cannot be combined with negative auto-feedback at perception level (and lower), entailing negative auto-feedback at interpretation level (and higher). In case two dialogue acts are in conflict with each other, the evaluation agent has to cancel one of them on the basis of some priority ordering among the candidates.

The occurrence of conflicting dialogue acts implies that the context model allows inconsistencies. Although this is undesirable and should be avoided in the design of the dialogue manager, it is nevertheless preferable to design the evaluation agent in such a way that it can deal with *any* combination of dialogue acts, irrespective of how the candidates were generated on the basis of the context model. Moreover, the context model does allow for inconsistencies between new information derived from input dialogue acts which has not yet been integrated into the context model. During the interpretation phase, this new information is stored in the *pending context* and ends up in the main context model only after successful evaluation. The detection of an inconsistency between the pending and the definitive context therefore results in an evaluation level processing problem, triggering the auto-feedback agent to generate a corresponding negative feedback act.

6.2 Pragmatic Constraints

Pragmatically speaking, two acts A_1 and A_2 are inconsistent if there is an implicated function A_1' of A_1 which has preconditions that are inconsistent with those of A_2 or with the preconditions of an implicated function A_2' of A_2. Questions and requests for example implicate that the speaker wants to release the current turn, hence the speaker does not want to have the next turn himself. Such acts as Stalling or Pausing, but also Self-correction, Error-signalling and Retraction implicate that the speaker wants to keep the turn himself. Corrections and Completions implicate positive feedback at the level of evaluation and hence cannot be combined with a positive auto-feedback act at interpretation level or lower, since these implicate negative feedback at the level of interpretation. A Contact Check carries an implication of negative perception of the addressee's linguistic or non-verbal behaviour, and can therefore not be combined with Opening, which carries an implication of positive perception of the addressee's behaviour. Similarly to logical constraints, in the case of two conflicting dialogue act candidates the evaluation agent will cancel the candidate with the lowest priority.

6.3 Constraints for Segment Sequences

For sequential multifunctionality within turns there are fewer and softer constraints on dialogue act combination than for simultaneous multifunctionality. The combination of two mutually exclusive acts in a sequence is in principle possible. A speaker who wants to construct coherent and logically consistent turns should not combine logically or pragmatically conflicting dialogue acts associated with segments within the same turn; however, such combinations cannot be excluded entirely, since a speaker can perform a dialogue act by mistake and subsequently correct himself. Obviously, in the design of a dialogue system this is mostly a phenomenon that is relevant for user input interpretation and cannot realistically be considered relevant for dialogue act generation.

6.4 Constraints Defining Dialogue Strategies

Given a list of dialogue act candidates that have no conflicts arising from either entailments or implicatures, there might still be reasons to put further constraints on combining these acts. The choice of such additional conditions depends on the particular settings the dialogue system is used in, and in fact offers a way to implement different dialogue strategies and styles of communication. In the next subsections several cases will be illustrated where the generation of multiple dialogue act candidates offers such options.

6.4.1 Negative Auto-feedback

As seen above, combinations of answers and negative auto-feedback on the level of either perception or interpretation give a logical conflict, but combinations of answers and negative feedback on the level of evaluation do not. The Task Agent can be triggered by a new user goal, even if this is part of the pending context only. This would be the case if the dialogue act recogniser detected a domain question in the user utterance, but the context manager did not yet check this new information for consistency with the context model, or already detected an inconsistency (i.e., an evaluation problem was encountered). As a result, the candidates list could contain both an answer to the question and a negative auto-feedback act on the level of evaluation, and constraining the combination of these acts is a strategic matter.

The dialogue fragment in (14) illustrates such a situation: after processing U2, the system detects a conflict between the user knowing where the send button is (from U1) and wanting to know where it is (from U2). This results in the generation of a negative auto-feedback act, and at the same time an answer to the question in U2. In generating system utterance S2 only the feedback act was selected, whereas alternatively both acts could have been selected, resulting in S2'. Which system response is the best is a strategic matter and depends on the global dialogue conditions as well as local conditions such as confidence scores propagated from an understanding module into the context model. For example, if the system was relatively confident about the interpretation of U1, S2' might be a more efficient choice than S2.

(14) *U1* : *I see the send button.*
S1 : *okay.*
U2 : *where is the send button?*

S2 : *but you just told me you saw the send button!*

S2' : *the send button is on the bottom right, but you just told me you saw it!*

In the case of S2, the answer to U2 is not cancelled but postponed until it is clear whether the system had misinterpreted U1 or U2. Consider the continuation of (14) in fragment (15): if it turns out that the system misinterpreted U1, as revealed in U3, the answer to the correctly interpreted question in U2 can be generated, resulting in S3; alternatively, if the system misinterpreted U2, the answer can be cancelled and replaced by an answer to the corrected question in U3'. In order to do this, the evaluation agent uses the updated processing status of U1 and U2 in the context model.

(15) *U1* : *I see the send button.*
S1 : *okay.*
U2 : *where is the send button?*
S2 : *but you just told me that you saw the send button!*

U3 : *no, I told you that I* needed *it.*
S3 : *oh, hold on ... the send button is on the bottom right.*

U3' : *no, I wanted to know where the* print *button is.*
S3' : *oh, hold on the print button is on the bottom left.*

6.4.2 Negative Allo-feedback

In the example dialogue fragment in (16), user and system are discussing a music concert by the Borodin Quartet. The system asks a question S1 and the user responds with a return question U1 which, to the system, seems unrelated. After processing U1, expecting some answer in the form of numerical information, the system may conclude that he misinterpreted the user, resulting in a negative auto-feedback act produced by the auto-feedback agent. The allo-feedback agent may also produce a negative feedback act on account that the user misinterpreted S1. Finally, the task agent might construct an answer to the question in U1 and also reproduce an act to get to know the number of tickets the user wants, as attempted earlier in S1. Strategic considerations are at the basis of the constraints that determine which of these candidates are selected. In S2, priority is given to the negative auto-feedback act; in S2a, the allo-feedback act is favoured; in S2b the answer to U1 is given; in S2c finally, the allo-feedback and answer acts are both realised.

(16) *S1* : *how many tickets do you want?*
U1 : *how much is the Kronos Quartet concert?*

S2 : *Sorry, I do not understand what you mean.*

S2a : *No, I would like to know the number of tickets you want*

S2c : *The Kronos Quartet concert is 30 euro.*

S2b : *The Kronos Quartet concert is 30 euro, but I asked about the Borodin Quartet.*

Again, the constraints used to make the choice between the presented options depend on the global dialogue settings and can also be conditional to the local context.

6.4.3 Scheduling Task Acts

Another type of strategic consideration is related to the planning of task acts and does not involve the combination of dialogue acts from different dimensions. For example, if the system has several questions to ask to the user, it has to decide whether to combine these questions in a single turn, or to take several turns to collect the information. The latter strategy is the more conservative one, and is used in situations where the risk of misunderstandings is higher, like in noisy environments or in general where the quality of speech recognition is limited.

On the basis of the user's input, the generation of several task-oriented dialogue acts can be triggered at once. Some user question or request could trigger several questions the system needs the user to answer before he can answer the question or carry out the request. In the case of several task-oriented dialogue acts, the relative priorities of these candidates are based on task-specific considerations. This could be based on some preferred, logical order in which subtasks should be carried out; in route-planning for example, it might be preferable to ask for the destination location before asking for the date on which the user wants to travel. The constraints defining the optimal scheduling could also be trained from data, in analogy to the reinforcement learning approaches that have been successfully applied in dialogue systems research.

6.4.4 Positive Auto-feedback

Another strategic issue involves the choice of whether or not to explicitly produce a dialogue act that is already implicated by another candidate. For example, a positive auto-feedback act does not need to be generated explicitly, if there is an answer candidate available as well. However, in certain settings, there might be good reasons for explicitly performing the implied dialogue act anyway. Every time the system reaches some level of successful processing of a user utterance, the auto-feedback agent may be triggered to produce a candidate dialogue act signalling this to the user. However, actually generating this dialogue act in all of these cases leads to a kind of communicative behaviour that can be experienced by the user as rather annoying. Instead, positive feedback should be generated only occasionally, depending on the specific communicative setting and the style of communication that is considered appropriate for that setting. In the case of dialogues involving the transfer of important information such as credit card numbers, it is desirable to give more positive feedback, but in the case of more informal dialogues, too much positive feedback should be avoided.

In the case of a train time table information system, giving positive feedback can be a good strategy. For example, in the dialogue fragment (17), after successfully processing U2, the system has gathered enough information from the user in order to answer the original question U1. In S2, the system generates this answer, where positive feedback about understanding U2 is implicated. The system might also give this feedback explicitly, resulting in the alternative responses in S2' and S2" (illustrating the well-known strategy of implicit verification).

(17) *U1* : *I'd like to know when the next train to Amsterdam is leaving.*
S1 : *From where are you travelling?*
U2 : *From Tilburg.*

S2 : *The next train leaves at 10:30h from platform 1.*

S2' : *So you want to go from Tilburg to Amsterdam.*
The next train leaves at 10:30h from platform 1.

S2" : *The next train from Tilburg to Amsterdam leaves at 10:30h from platform 1.*

6.4.5 Social Obligations Management

The role of social obligations management acts in the process of selecting and combining dialogue acts is also a matter of strategy and choosing an appropriate style of communication. In some settings there is less need for task efficiency and it is more appropriate for the system to behave more socially. In such cases social acts such as thanks and apologies, as well as their response counterparts, can be included in the system contributions with greater frequency. Apologies can be used typically in combination with negative feedback acts (*'I'm sorry, I did not hear what you were saying'*), whereas greetings are typically used in combination with contact management (*'Hello?'*).

6.5 Evaluation Agent Design

Having discussed the various types of constraints for combining dialogue acts, and the logical, pragmatic, strategic and stylistic considerations involved, a procedure for the operation of the evaluation agent can be formulated that consists of three phases:

1. In the first phase, the dialogue act candidates are inspected for any conflicts, which are subsequently resolved by cancelling lower-priority acts.

2. In the second phase, the remaining list of non-conflicting candidates is evaluated from a pragmatic and dialogue strategic point of view, possibly resulting in cancelling or postponing some of the dialogue act candidates.
3. Finally, in the third phase, combinations of dialogue acts are selected that can actually be realised in multifunctional system utterances. Some combinations of dialogue acts may not carry any logical conflicts, but the particular natural language may not provide a multifunctional utterance for the dialogue acts to be realised. For example, a question in one dimension cannot be combined with an information in another dimension using one single utterance, because the question requires an interrogative, and the information, a declarative sentence.

Besides the construction of multifunctional utterances, some of the dialogue acts can also be realised in a non-verbal manner, for example by means of animations on the graphical user interface of the system, or by means of gestures made by the system if it is an embodied virtual agent.

7 Conclusion

In this chapter a new approach to dialogue management has been discussed in which the multidimensionality of communication is reflected. Using the semantic framework of Dynamic Interpretation Theory, the notion of multifuntionality was described and empirically motivated. A general dialogue manager architecture was outlined, involving multiple dialogue act agents, each of which focused on one of the dimensions of communication. An initial proof-of-concept implementation of this dialogue manager was realised and integrated in the IMIX interactive Question Answering system.

An important element in the dialogue act generation process is dealing with the interdependencies that may exist between dialogue acts addressing different dimensions. It is the task of the evaluation agent to take these dependencies into account when combining the candidate dialogue acts as produced by the dialogue act agents. Giving priority to some dialogue acts and postponing or cancelling others was shown to involve logical, strategic and pragmatic considerations, besides specific language generation issues that were not discussed. It was also shown how the process of evaluating candidate dialogue acts allows for implementing different dialogue strategies and communication styles in the dialogue manager.

The main focus of future research is to further develop the empirical and theoretical analysis of multifunctionality and further specify constraints on the process of dialogue act generation. The aim is to make different possible dialogue strategies and styles of communication explicit in these specifications. The implementation of the resulting context update and generation mechanisms will also provide the basis for a framework for evaluating these notions on real (corpus) data.

References

Op den Akker R, Bunt H, Keizer S, van Schooten B (2005) From question answering to spoken dialogue: Towards an information search assistant for interactive multimodal information extraction. In: Proceedings of the 9th European Conference on Speech Communication and Technology, Interspeech 2005, pp 2793–2796

Allen J, Core M (1997) DAMSL: Dialogue Act Markup in Several Layers (Draft 2.1). Technical Report. University of Rochester, Rochester, NY

Allwood J (2000) An activity-based approach to pragmatics. In: Bunt H, Black W (eds) Abduction, Belief and Context in Dialogue, John Benjamins, Amsterdam, pp 47–81

Bunt H (1989) Information dialogues as communicative action in relation to partner modelling and information processing. In: Taylor M, Néel F, Bouwhuis D (eds) The structure of multimodal dialogue, North-Holland, Amsterdam, pp 47–74

Bunt H (1995) DIT – Dynamic interpretation and dialogue theory. In: Taylor M, Néel F, Bouwhuis D (eds) The structure of multimodal dialogue, Volume 2, John Benjamins, Amsterdam, pp 139–166

Bunt H (2000) Dialogue pragmatics and context modelling. In: Bunt H, Black W (eds) Abduction, Belief, and Context in Dialogue, John Benjamins, Amsterdam, pp 81–150

Bunt H (2009a) A framework for dialogue act specification. In: Heylen D, Pelachaud C, Catizone R, Traum D (eds) Proceedings of EDAML–AAMAS Workshop "Towards a Standard Markup Language for Embodied Dialogue Acts", Budapest, pp 36–36

Bunt H (2009b) Multifunctionality and muldimensional dialogue semantics. In: Proceedings of DiaHolmia, 13th Workshop on the Semantics and Pragmatics of Dialogue, Stockholm, pp 3–14

Bunt H (2009c) Semantic annotations as complementary to underspecified semantic representations. In: Proceedings of IWCS-8, the 8th International Workshop on computational Semantics, Tilburg, pp 33–44

Bunt H (2010a) Multifunctionality in dialogue and its interpretation. Computer Speech and Language, Special issue on dialogue modeling, Y Wilks, editor

Bunt H (2010b) Multifunctionality in dialogue and its interpretation. In: Toward Autonomous, Adaptive, and Context-Aware Multimedia IInterfaces, Anna Esposito, editor, Berlin: Springer

Bunt H, Keizer S (2005) Dialogue semantics links annotation to context representation. In: In Joint TALK/AMI Workshop on Standards for Multimodal Dialogue Context, Edinburgh, URL `http://homepages.inf.ed.ac.uk/olemon/standcon-SOI.html`

Bunt H, Morante R, Keizer S (2007) An empirically based computational model of grounding in dialogue. In: Proceedings of the 8th SIGdial Workshop on Discourse and Dialogue, Antwerp, Belgium, pp 283–290

Bunt H, Alexandersson J, Carletta J, Choe JW, Fang A, Hasida K, Lee K, Petukhova V, Popescu-Belis A, Romary L, Soria C, Traum D (2010) Towards an ISO standard for dialogue act formal annotation. In: Proceedings 7th International Conference on Language Resources and Evaluation (LREC 2010), Malta, Paris: ELRA

Ginzburg J (1997) Resolving Questions I & II. Linguistics and Philosophy 17; 18

Grice H (1975) Logic and conversation. In: Cole P, Morgan J (eds) Syntax and Semantics, Vol.3: Speech Acts, Academic Press, New York, pp 43–58

Keizer S, Bunt H (2006) Multidimensional dialogue management. In: Proceedings of the SIGdial Workshop on Discourse and Dialogue, Sydney, Australia, pp 37–45

Keizer S, Bunt H (2007) Evaluating combinations of dialogue acts for generation. In: Proceedings of the SIGdial Workshop on Discourse and Dialogue, Antwerp, Belgium, pp 158–165

Keizer S, Morante R (2006) Context specification and update mechanisms for dialogue management. In: Proceedings of the 18th BeNeLux Conference on Artificial Intelligence (BNAIC'06), Namur, Belgium, pp 181–188

Larsson S (1998) Questions Under Discussion and Dialogue Moves. In: Proceedings of the Twente Workshop on the Formal Semantics and Pragmatics of Dialogue, Twente, pp 209–247

Petukhova V, Bunt H (2009a) Dimensions in communication, URL `http://www.tilburguniversity.nl/faculties/humanities/ticc/research/technicalreports/`, TiCC Technical Report TR 2009-002, Tilburg University

Petukhova V, Bunt H (2009b) The independence of dimensions in multidimensional dialogue act annotation. In: Proceedings NAACL HLT Conference, Boulder, Colorado

Petukhova V, Bunt H (2010) introducing communicative function qualifiers. In: Proceedings Second International Conference on Global Interoperability for Language resources (ICGL-2), Hong Kong, pp 123–131

Poesio M, Traum D (1997) Conversational Actions and Discourse Situations. Computational Intelligence 13 (3):309–347

Poesio M, Traum D (1998) Towards an Axiomatization of Dialogue Acts. In: Proceedings of the Twente Workshop on the Formal Semantics and Pragmatics of Dialogue, Twente, pp 309–347

Stent A (2002) A conversation acts model for generating spoken dialogue contributions. Computer Speech and Language, Special Issue on Spoken Language Generation 16(3–4):313–352

Traum D, Larsson S (2003) The information state approach to dialogue management. In: van Kuppevelt J, Smith R (eds) Current and New Directions in Discourse and Dialogue, Kluwer, Dordrecht, pp 325–354

Turunen M, Hakulinen J, Räihä KJ, Salonen EP, Kainulainen A, Prusi P (2005) An architecture and applications for speech-based accessibility systems. IBM Systems Journal 44(3):485–504

Part III
Fusing Text, Speech, and Images

Experiments in Multimodal Information Presentation

Charlotte van Hooijdonk, Wauter Bosma, Emiel Krahmer, Alfons Maes and Mariët Theune

Abstract An important research question in multimodal information presentation is, which presentation modes, such as text, speech, images and animations, or combinations thereof, are most suitable for meeting different communicative goals (e.g., instruct, inform, and persuade)? This chapter describes three experiments carried out to find answers to this question. Experiment 1 investigated how people produced (multimodal) information presentations in the medical domain. In Experiment 2, people were asked to evaluate these information presentations on their informativeness and attractiveness. Finally, Experiment 3 evaluated two different methods (caption-based selection versus section-based selection) to automatically illustrate answers to medical questions, and compared the results to those of manually created presentations.

1 Introduction

Recent developments in computer technology have led to new possibilities for presenting information and for a renewed interest in the effects of different

---

Charlotte van Hooijdonk
VU University Amsterdam, Amsterdam, The Netherlands, e-mail: cmj.van.hooijdonk@let.vu.nl

Wauter Bosma
VU University Amsterdam, Amsterdam, The Netherlands, e-mail: w.bosma@let.vu.nl

Emiel Krahmer
Tilburg University, Tilburg, The Netherlands, e-mail: e.j.krahmer@uvt.nl

Alfons Maes
Tilburg University, Tilburg, The Netherlands, e-mail: maes@uvt.nl

Mariët Theune
University of Twente, Enschede, The Netherlands, e-mail: m.theune@ewi.utwente.nl

presentation modes. Naturally, this raises questions, such as "Which presentation modes are most suitable given a particular communicative goal?" and "How should different presentation modes be combined?" The IMOGEN (Interactive Multimodal Output GENeration) project addressed these questions. This project was embedded in the Dutch national research programme IMIX (Interactive Multimodal Information eXtraction). Within IMIX a multimodal medical Question Answering (QA) system was developed. The purpose of this system is to answer encyclopedic medical questions from non-expert users. Questions can be typed or spoken (in Dutch), and answers are presented using speech, text and pictures. Questions can be asked in isolation, but the system is also capable of engaging in dialogues and answering follow-up questions.

In the IMOGEN project different aspects of multimodal information presentation were studied in order to improve the output quality of the Question Answering (QA) systems. Early research in the field of QA concentrated on answering factoid questions, i.e. questions that have one word or phrase as their answer, such as "Amsterdam" in response to the question "What is the capital of the Netherlands?" The presentation mode of the answers to these questions was typically text only. Nowadays, QA systems are also expected to give answers to more complex questions, which might be more informative and effective if they contained multiple presentation modes, such as text and a picture. Here the focus is on questions in the medical domain, since the QA system developed as a demonstrator in IMIX aimed at answering encyclopedic medical questions from non-expert users.

People can have different medical questions, including factoid definition questions, such as "What is RSI?" and procedural questions about how to take care of one's health, such as "How to prevent RSI?" People may also have different information needs. In some situations, they are satisfied with a short answer in which, for example, the abbreviation RSI is clarified (*Repetitive Strain Injury*). In other cases, longer answers are wanted in which more information is given about the causes and consequences of the disorder. (For example, *RSI stands for Repetitive Strain Injury. This disorder involves damage to muscles, tendons and nerves caused by overuse or misuse, and affect the hands, wrists, elbows, arms, shoulders, back, or neck.*) The answers to these medical questions can be presented through text or through a combination of presentation modes, such as text and a static or dynamic picture. For example, the most suitable answer presentation to the definition (*what*) question "What does RSI stand for?" would probably be a short textual answer, such as "RSI stands for Repetitive Strain Injury". The answer to the procedural (*how*) question "How to organise a workspace in order to prevent RSI?" would probably be more informative if it contained a picture. This raises the question of how to determine for a given question, whether a short or a long answer would be preferable and which (combinations of) presentation modes are most suitable.

Multimodal information presentation has been studied in various research fields with various outcomes. Research in cognitive and educational psychology focused on how multimodal presentations affect the users' understanding, recall and processing efficiency of the presented material, e.g., (Carney and Levin, 2002; Mayer, 2005; Tversky et al, 2002). Guidelines resulting from this research often relate to specific types of information used in specific domains, for example cause and effect chains (Mayer and Moreno, 2002) or procedural information (Michas and Berry, 2000). Yet, these guidelines do not tell us which modalities are most suited for which information types, as each learning domain has its own characteristics (van Hooijdonk and Krahmer, 2008).

Research in user interfaces has tried to classify and characterise different presentation modes. For example, Bernsen (1994) proposed a taxonomy of generic unimodalities consisting of various features. Other scholars studied the so-called *media allocation problem* (i.e., how to determine which information to allocate to which medium) and tried to identify which factors play a role in media allocation (Arens et al, 1993). They found out that many factors are relevant: the nature of the information, the communicative situation, the goals of the producer, and the features of the addressee.

In short, attempts have been made to generate optimal multimodal information presentations resulting in several presentation mode guidelines, frameworks, and taxonomies. What is still needed is information about people's modality preferences in producing and evaluating presentations. Therefore, three experiments were carried out following the approach of Heiser et al (2004). The experiments investigated multimodal information presentation in the context of a medical QA system. In Experiment 1:- people were asked to produce information presentations, which were then rated by others in Experiment 2. In Experiment 3:- the answer presentations manually produced in Experiment 1 were compared to presentations with automatically retrieved pictures.

This chapter presents the three experiments. In Experiment 1 we wanted to know how non-experts design (multimodal) answers to medical questions, distinguishing between *what* questions and *how* questions. In Experiment 2 we concentrated on how people evaluate multimodal (text+picture) answer presentations on their informativeness and attractiveness. In Experiment 3 we evaluated two versions of an automatic picture selection method, and compared answer presentations with automatically selected pictures to answer presentations with manually selected pictures.

2 Experiment 1: Production of Multimodal Answers

This section presents an experiment that was carried out to determine which modalities people choose to answer different types of questions. In the experiment, participants had to create (multimodal) presentations of answers to general medical questions. More details on the experiment can be found in van Hooijdonk et al (2007a).

2.1 Participants

Participants were 111 students of Tilburg University, who participated for course credits (65 female and 46 male). Their average age was 22 (SD = 2.10, min. = 19, max. = 32). All participants were native speakers of Dutch. All were second-year undergraduate students, who had received Internet search training in the first year of their studies. They were all familiar with Microsoft PowerPoint, and used it on a regular basis (daily: 3.6%, weekly: 22.5%, monthly: 51.4%, yearly: 18.0%, never: 4.5%). Finally, participants indicated on one 7-point semantic differential that their PowerPoint skills were above average (M = 5.01, SD = 1.10).

2.2 Stimuli

Participants were given one of four sets of eight general medical questions for which the answers could be found on the Internet. They had to provide two types of answers per question, a short and a long answer, using whatever combination of presentation modes they wanted. They did not get explicit instructions on the number of words or pictures to be used in their answers. Participants were specifically asked to present the answers as they themselves would prefer to find them in a QA system. Questions and answers had to be presented in a fixed format in PowerPoint™ with areas for the question ('vraag') and the answer ('antwoord'). Participants were given a short introduction about PowerPoint in which they were acquainted with inserting different types of objects into PowerPoint. Also, they received a PowerPoint manual. Of the eight questions in each set, four were randomly chosen from one hundred medical questions formulated to test the IMIX system. Of the remaining four questions, two were *what* questions (e.g., "What are thrombolytic drugs?") and two were *how* questions (e.g., "How to apply a sling to the left arm?").

2.3 Coding System and Procedure

Each answer was coded on the presence of visual media (i.e. photos, graphics, and animations) – pictures, in short – and on the function of these pictures in relation to the text, loosely based on Carney and Levin (2002), i.e., decorative, representational, or informative.

Decorative function A picture has a decorative function if removing it from the answer presentation does not alter the informativeness of the answer in any way. Figure 1 shows an example of an answer with a decorative picture. The answer to the question "What are the side effects of a vaccination for diphtheria, whooping cough, tetanus, and polio?" consists of a combination of text and a graphic.

The text describes the side effects of the vaccination, while the graphic shows a syringe. The answer would not be less informative if the graphic was absent.

Representational function A picture has a representational function if removing it from the answer presentation does not alter the informativeness of the answer, but its presence clarifies the text. Figure 2 shows an example of an answer presentation with a representational picture. The question "What types of colitis can be distinguished?" is answered through text and a graphic. The text describes the four types of colitis and where they are located in the intestines. This information is visualised in the graphic.

Informative function A picture has an informative function if removing it from the answer presentation decreases the informativeness of the answer. If an answer only consists of a picture, it automatically has an informative function. Figure 3 shows an example of an answer with an informative picture. The answer to the question: "How can I strengthen my abdominal muscles?" consists of text and photos. The text describes some general information about abdominal exercises (i.e., an exercise programme should be well balanced and train all abdominal muscles). The last sentence refers to four exercises that can be done to strengthen the abdominal muscles. These exercises are illustrated by eight photos. For each exercise two photos are given, indicating the first (a) and last (b) step of the exercise.

In total 1776 answers were collected (111 participants $\times$ 8 questions $\times$ 2 answers). One of the participants omitted one answer, so that the final data set consisted of 1775 answers. Six analysts independently coded the same set of 111 answers. Subsequently, every analyst independently coded a part of the total corpus (approximately 300 answers). Calculations of Cohen's κ showed that the analysts agreed almost exactly in their judgement of the occurrence of photos ($\kappa = .81$), graphics ($\kappa = .83$), and animations ($\kappa = .92$). An almost perfect agreement was also reached in assigning the function of the picture media ($\kappa = .83$).

2.4 Results

Analysis of the complete corpus of coded answer presentations showed that almost one in four answers contained one or more pictures ($n = 442$), consisting of graphics ($n = 232$), photographs ($n = 124$), or animations ($n = 49$). In 37 cases, a combination of these media was used.

Answer length

Long answers (M = 86, SD = 60) contained significantly more words than short answers (M = 18, SD = 25), $t(168.78) = -10.58$, $p < .001$ (since Levene's test was significant, a correction on the degrees of freedom was made). Table 1 shows that long answers contained significantly more pictures than short answers ($\chi^2(1) = 173.89$, $p < .001$). Moreover, the distribution of the functions of visual media differed significantly over answer length ($\chi^2(2) = 33.79$, $p < .001$). Decorative pictures occurred most often in short answers ($\chi^2(1) = 4.07$, $p < .05$), whereas representational pictures occurred most often in long answers ($\chi^2(1) = 125.78$, $p < .001$). Informative pictures occurred most often in short answers ($\chi^2(1) = 23.81$, $p < .001$).

Table 1 Percentages of function of visual media related to short and long answers ($n = 442$).

	Short answers ($n = 101$)	Long answers ($n = 341$)
Decorative pictures ($n = 70$)	26.7	12.6
Representational pictures ($n = 201$)	20.8	52.8
Informative pictures ($n = 171$)	52.5	34.6

Question type

Analysis of the two *what* questions and the two *how* questions ($n = 887$, of which 271 contained pictures) showed that pictures occurred significantly more often in *how* questions ($\chi^2(1) = 29.23$, $p < .001$). Table 2 also shows that answers to *what* questions contained significantly more decorative and representational pictures,

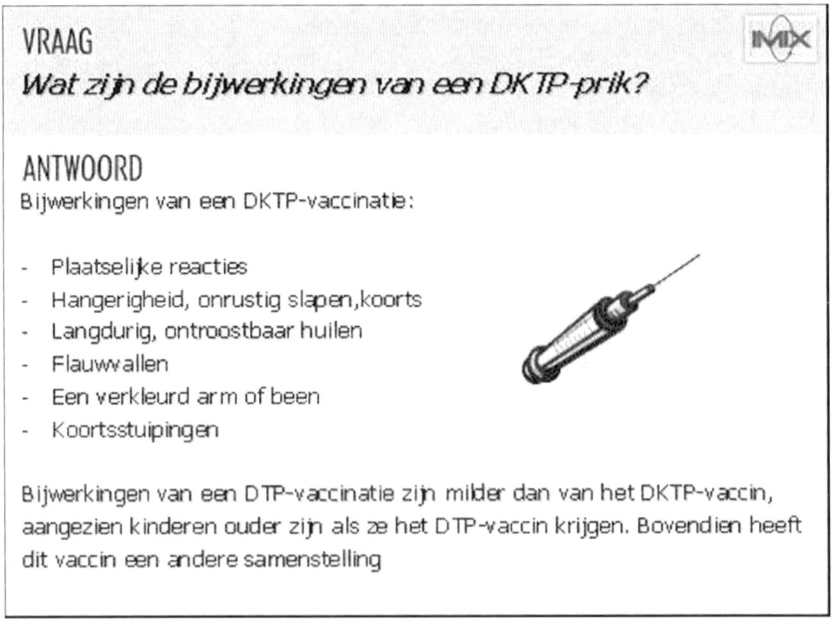

Fig. 1 Example of an answer with a decorative picture.

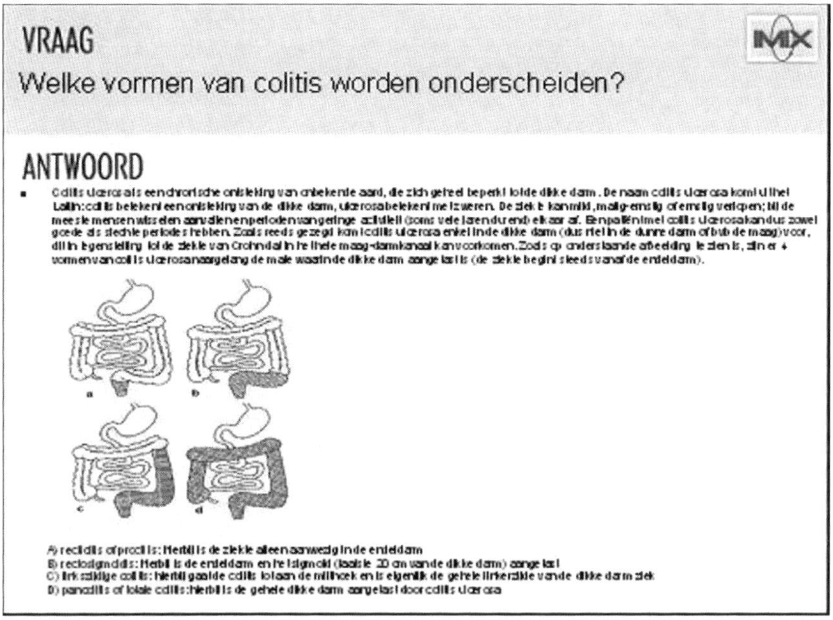

Fig. 2 Example of an answer with a representational picture.

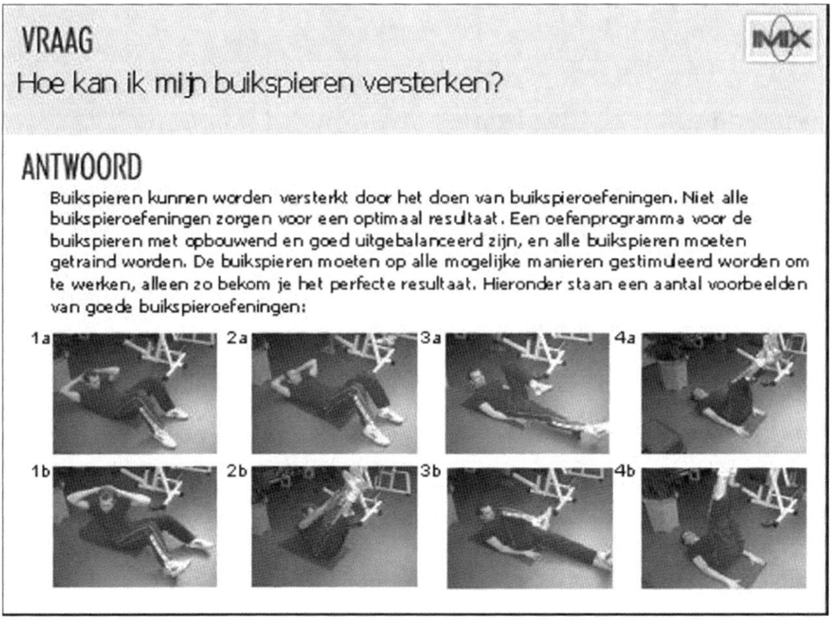

Fig. 3 Example of an answer with an informative picture.

while answers to *how* questions contained more informative pictures ($\chi^2(2) = 22.70$, $p < .001$).

Table 2 Percentages of functions of pictures related to *what* questions and *how* questions ($n = 271$).

	What questions ($n = 91$)	*How* questions ($n = 180$)
Decorative pictures ($n = 27$)	19.8	5.0
Representational pictures ($n = 129$)	53.8	44.4
Informative pictures ($n = 115$)	26.4	50.6

2.5 Conclusion

The results showed that people made use of multiple presentation modes in their answers and that the design of these presentations was affected by answer length and question type. What is not clear, is how people evaluate multimodal (text+picture) answer presentations. In the next section, an evaluation experiment is discussed in which this issue was investigated.

3 Experiment 2: Evaluation of Multimodal Answers

This section presents Experiment 2, which was conducted to investigate how users evaluate different types of multimodal answer presentations. In this experiment, participants had to assess the informativeness and attractiveness of answer presentations for different types of medical questions. More details on the experiment can be found in van Hooijdonk et al (2007b).

3.1 Participants

The participants were 108 native speakers of Dutch (66 female and 42 male). Their average age was 25 (SD = 8.24, min. = 18, max. = 64). None had participated in Experiment 1.

3.2 Design

The experiment had 16 (question) $\times$ 2 (short or long answer) $\times$ 3 (decorative picture, informative picture, or no picture) mixed factorial design, with the question as a within participants variable, and the answer length and picture type as a between participants variable. The dependent variables were the participants' assessment of: (a) the clarity of the text, (b) the informativeness of the answer presentation, (c) the attractiveness of the answer presentation, (d) informativeness of the text-picture combination and (e) the attractiveness of the text-picture combination. The participants were randomly assigned to an experimental condition.

3.3 Stimuli

16 medical questions were selected, for which the corpus collected in Experiment 1 contained: (i) an informative picture, which added new information to the answer and (ii) a decorative picture, which did not. Although the results of Experiment 1 showed that participants used different picture types when producing short and long answers, only informative and decorative pictures were taken into account in Experiment 2. The hypothesis was that decorative pictures would be evaluated as most attractive but least informative, as they might make the text more vivid but do not add new information to the textual answer. Informative pictures on the other hand would be evaluated as least attractive but most informative, as they add new information to the textual answer but do not necessarily make the information more attractive. Representational pictures visually display the main topic of the textual answer, but do not add new information. In this respect, they are quite similar to decorative pictures, as they might make the textual answer more vivid but add no new information to the textual answer. Therefore, representational and decorative pictures were combined into the category of decorative pictures.

The set of selected questions consisted of eight *what* questions and eight *how* questions. For each question a short and a long textual answer was formulated. The textual answers were chosen from the set of answers collected in Experiment 1. Small adjustments were made to these answers in order to make them more comparable. The short answer gave a direct answer to the question, while the long answer also provided some relevant background information. The average length of the short answers was 26 words and the average length of the long answers was 66 words. It was made sure that the type of question did not affect the answer length for short answers ($F[1,14] = 3.59$, $p = .08$), nor for long answers ($F < 1$).

Answers to the medical questions were presented in six different presentation formats: a short and a long textual answer, each used (i) on its own (unimodal), (ii) combined with an informative picture (multimodal) and (iii) combined with a decorative picture (multimodal). The remainder of this section only discusses the multimodal answer presentations.

Two multimodal answer presentations, a short and a long answer, contained a decorative picture. Figure 4 shows the short and the long answer to the question "How to organise a workspace in order to prevent RSI?", illustrated with a decorative photograph showing a workspace. The other two multimodal answer presentations contained an informative picture. Figure 5 shows the short and the long answer to the same question as in Figure 4, but this time illustrated with an informative graphic. The graphic depicts an ergonomic workspace in detail. It should be noted that all answer presentations were designed in such a way that the textual element by itself already contained enough information to answer the question; the informative pictures only added relevant background information.

All answer presentations were presented to the participants in a random order and this was the same for all participants.

3.4 Procedure

The experiment was conducted using WWSTIM (Veenker, 2005), a CGI-based script that automatically presents stimuli to the participants and transfers all data to a database. This enabled the experiment to be run via the Internet.

The participants received an e-mail inviting them to take part in the experiment. This e-mail briefly stated the goal of the experiment, the amount of time it would take to participate, the possibility of winning a gift voucher, and the URL of the experiment. When accessing the website for the experiment, participants received instructions about the procedure. Next, they entered their personal data (i.e. age, gender, level of education, and optionally their e-mail to win a gift voucher). After a short practice session, participants studied 16 question-answer combinations, one at a time. After each combination, they were shown the same combination with at the bottom five seven-point semantic differentials (implemented as radio buttons) which they had to use to rate the informativeness of the answer (the answer presentation is informative/ not informative), the attractiveness of the answer (the answer presentation is attractive/ not attractive), the informativeness of the text-picture combination (the text-picture combination is informative/ not informative), the attractiveness of the text-picture combination (the text-picture combination is attractive/ not attractive), and the clarity of the text (the text is formulated in a simple/ complex way).

3.5 Results

We report only on the participants' assessment of the informativeness and the attractiveness of the text-picture combinations. For (partial) results on the other presentation aspects evaluated by the participants, see Section 5.6, where they are compared to the results of automatically illustrated presentations.

Fig. 4 Examples of a short textual answer (top) and a long textual answer (bottom) with a decorative picture.

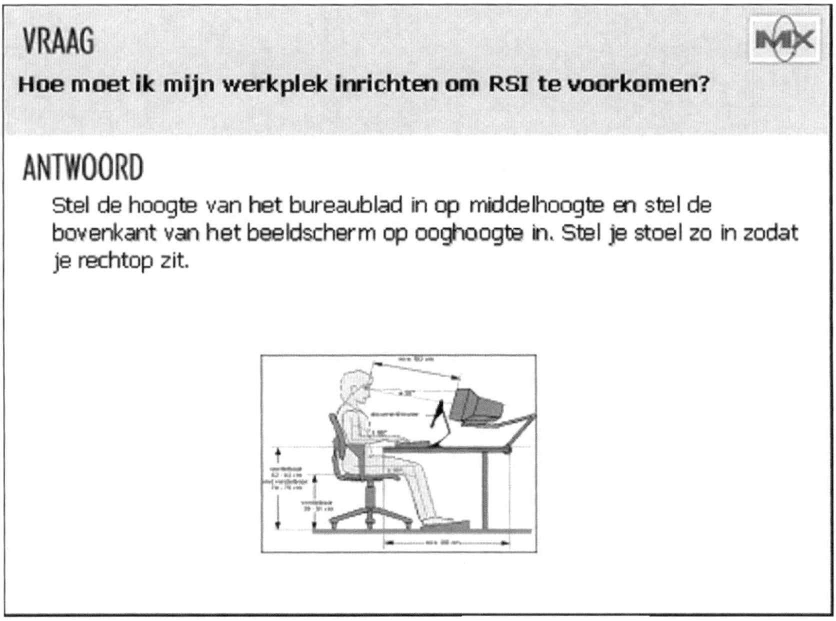

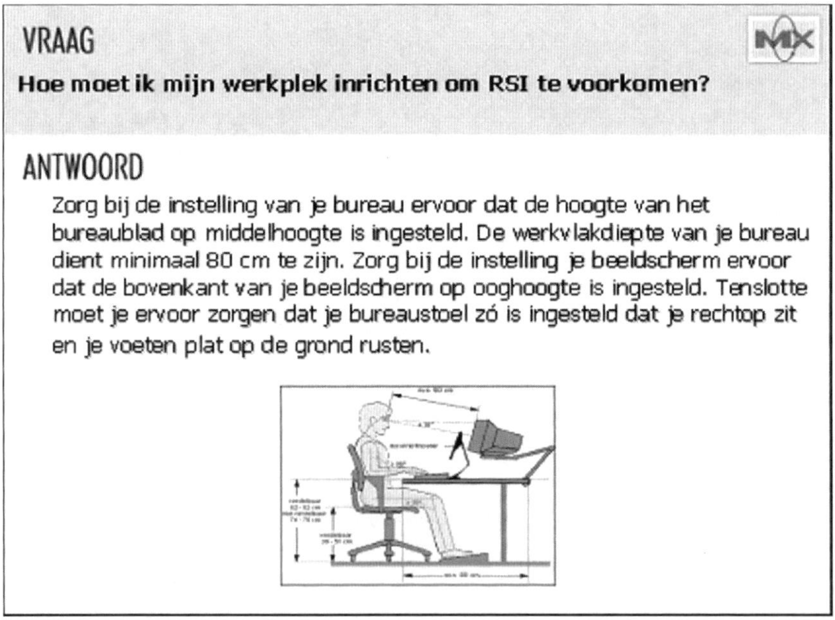

Fig. 5 Examples of a short textual answer (top) and a long textual answer (bottom) with an informative picture.

The results were tested for significance using a 4 (answer presentation) × 2 (question type) repeated measures analysis of variance (ANOVA). As shown in Table 3, short answers with an informative picture were evaluated as most informative, and short answers with a decorative picture as least informative ($F[3,68] = 9.32$, $p < .001$, $\eta_p^2 = .29$). Answers to *how* questions were rated as more informative than answers to *what* questions ($F[1,68] = 15.13$, $p < .001$, $\eta_p^2 = .18$). Finally, a relationship was found between answer presentation and question type ($F[3,68] = 4.27$, $p < .01$, $\eta_p^2 = .16$): for both short ($F[1,17] = 17.12$, $p < .005$, $\eta_p^2 = .50$) and long ($F[1,17] = 7.31$, $p < .025$, $\eta_p^2 = .30$) answers with an informative picture, answers to *how* questions were evaluated as more informative than answers to *what* questions. For answers with a decorative picture no significant differences were found between the two question types.

Long answers with an informative picture were evaluated as most attractive, long answers with a decorative picture were evaluated as least attractive ($F[3,68] = 4.64$, $p < .01$, $\eta_p^2 = .17$). Answers to *how* questions were evaluated as more attractive than answers to *what* questions ($F[1,68] = 20.59$, $p < .001$, $\eta_p^2 = .23$). No relationship was found between answer presentation format and question type ($F < 1$).

Table 3 Mean results for the informativeness and attractiveness of answer presentation types (ratings range from 1 = "very negative" to 7 = "very positive"; standard deviations in parentheses).

		Text with decorative picture		Text with informative picture	
Factor	Question type	Short	Long	Short	Long
Informative	What	3.83 (1.13)	4.01 (1.30)	4.91 (.81)	4.97 (1.20)
	How	3.70 (1.26)	4.27 (1.18)	5.53 (.70)	5.40 (.84)
	Total	3.76 (1.16)	4.14 (1.19)	5.22 (.69)	5.18 (1.00)
Attractive	What	3.93 (.87)	3.76 (1.14)	4.43 (.88)	4.69 (1.01)
	How	4.18 (1.12)	4.18 (1.10)	4.95 (.84)	5.08 (.76)
	Total	4.06 (.96)	3.97 (1.07)	4.69 (.75)	4.89 (.79)

3.6 Conclusion

The results show that answers with an informative picture were evaluated as more informative than answers with a decorative picture, especially for short answers, which is consistent with the production experiment (Experiment 1). The information load of the textual answers could explain these results. Short answers contain less information than long ones. Therefore, an informative picture adds more information to short answers than to long answers, and is thus perceived as more informative in combination with a short answer. Also, answers to *how* questions with an informative picture were evaluated as more informative than answers to *what* questions. Arguably, the medical procedures – as they occurred in this experiment – lend themselves better to be visualized than definitions, because they have a dynamic and spatial character. Interestingly however, long answers with informative pictures were evaluated as most attractive, suggesting that users like complete information together with highly informative pictures.

4 Automatic Production of Multimodal Answers

The previous sections discussed how humans produce and evaluate multimodal answers. However, most existing QA systems present their answers in one presentation mode, i.e. text snippets retrieved from a document corpus. Pictures that occur in the corpus documents are generally ignored, since the text-oriented retrieval methods used in QA systems cannot deal with them. A method for extending the answers returned by a QA-system with appropriate pictures has been proposed in (Bosma, 2005). This section describes the picture selection method, and the next section presents a user evaluation (Experiment 3) in which the results of two variations of this method are compared with the manually created multimodal answer presentations used in Experiment 2.

4.1 Multimedia Summarization

The approach to generating multimodal answers to questions is essentially automatic multimedia summarisation, using established techniques from automatic text summarisation. Most text summarisation methods (used in the context of a QA system) are based on comparative analyses between the user's query and parts of the source document(s). Multimedia summarisation faces the difficulty that different media have different features and thus cannot be directly compared (e.g., the word "red" cannot be directly compared to the colour red). Analysing and converting media content to a semantic representation has been proposed as a solution for this problem (van Deemter and Power, 2003; Maybury and Merlino, 1997; Nagao et al, 2002; Petrushin, 2007). However, automatic analysis of media content is difficult and often unreliable. Manual annotation is an alternative which answers some of these objections, but this is very laborious. Another solution, which according to de Jong et al (2007) is often overlooked, is to use related linguistic content for analysis, instead of the media items themselves. If related text adequately describes a media item, text-based retrieval methods can be used to retrieve non-textual media.

Multimedia presentations were automatically generated as answers to medical questions by using a query-based summarisation framework, described in more detail in the chapter *Text-to-Text Generation for Question Answering* by Bosma, Marsi, Krahmer and Theune, this volume, in a multimedia setting. The query-based summarisation framework relies on a combination of one or more feature graphs representing the source documents. A content unit can be a unit of any medium, such as a text snippet or a picture. The graphs express relations between the documents' content units, and are constructed using content (e.g. cosine similarity, see the next section) or context (e.g. layout) to relate to content units. This way, content can be presented for which there is only indirect evidence of relevance. For instance, a sentence that is adjacent – and thus contextually related – to a sentence that is similar to the query may be included in the answer, even though it is only linked to the query indirectly. This concept may also be applied to multimedia. A picture can be related to a piece of text by using layout information. A straightforward indication of relatedness of text and visual content is when the text is the picture's caption, but the paragraph or section in which the picture is located may also be considered as related to the picture.

4.2 Automatic Picture Selection

In the IMIX system, the approach sketched above is used to select the best picture to illustrate a given textual answer to a medical question. To find this picture, the illustration system compares the text of the answer with the picture-associated text. The more similar the two text passages, the more likely it is that the picture is relevant. The picture-associated text is interpreted as a textual representation of the picture. This may be either the picture's caption or the paragraph (or section if no single paragraph could be related to the picture) in which the picture was found. The relevancy of a picture for the answer is calculated as:

$$R_{picture}(i,t) = cosim(t, text(i)) \quad (1)$$

Where $R_{picture}(i,t)$ is the relevancy of picture i to text t; and $text(i)$ is the text associated with picture i. The function $cosim(a,b)$ calculates the cosine similarity of a and b.

Cosine similarity is a way of determining lexical similarity of text passages. The idea behind cosine similarity is that a text's meaning is compiled from the meaning of its words. The measure cosine similarity between two passages was represented with both texts as a vector whose elements represent the contribution of a word to the meaning of the passage. Before measuring the cosine similarity, words are stemmed using Porter's stemmer (Porter, 1997). The cosine similarity is calculated as follows:

$$cosim(a,b) = \frac{\sum_{k=1}^{n} a_k \cdot b_k}{|a| \cdot |b|} \quad (2)$$

Where $cosim(a,b)$ is the similarity of passages a and b; n is the number of distinct words in the passages. Both passages are represented as a vector of length n, with a_k representing the contribution of word k to passage a. The denominator ensures that passage vectors are normalised by their lengths. The value $|a|$ is the length of passage vector a, measured as $\sqrt{\sum_{k=1}^{n} a_k^2}$.

Determining how much a particular word contributes to the meaning of a passage is called *term weighting*. The system used *tf* · *idf* term weighting, i.e. the contribution of a word to a passage is calculated as the word's occurrence frequency in the passage (term frequency, TF) multiplied by the word's inverse document frequency (IDF). IDF is a measure of how characteristic the word is for a passage. To measure the inverse document frequency a large set of passages was required. For this we used the passage vectors of picture-associated text for all pictures in a medical corpus (see Section 5.3), plus the passage vector of the answer text. A word occurring in few of these passages receives a high IDF value, because the low occurrence rate makes it descriptive of the few passages it appears in. Conversely, a word occurring in many passages receives a low IDF value. The contribution of word k to passage a is measured as follows:

$$a_k = tf_{a,k} \cdot idf_k \tag{3}$$

Where $tf_{a,k}$ is the number of occurrences of word k in passage a; and idf_k is the IDF value of word k. The IDF value is calculated as follows:

$$idf_k = log \frac{|D|}{|\{d \mid d \in D \wedge k \in d\}|} \tag{4}$$

Where $|D|$ is the number of passages in the corpus (i.e. the number of pictures plus one); and the denominator is the number of documents which contain the word k. The final answer presentation consists of the textual answer and the most relevant picture and its caption.

Figure 6 shows an example of an answer presentation containing an automatically selected picture. In this figure and in Figure 7 the answer presentation is embedded in the web interface used for Experiments 2 and 3, which was designed to replicate the 'look and feel' of a medical QA system.

5 Experiment 3: Evaluating Automatically Produced Multimodal Answers

An experiment was carried out to evaluate two variants of the previously described approach by automatically adding pictures to textual answers. The study was largely identical to Experiment 2, except that it used automatically retrieved pictures instead of manually selected ones. More details on the experiment can be found in (Bosma et al, 2008).

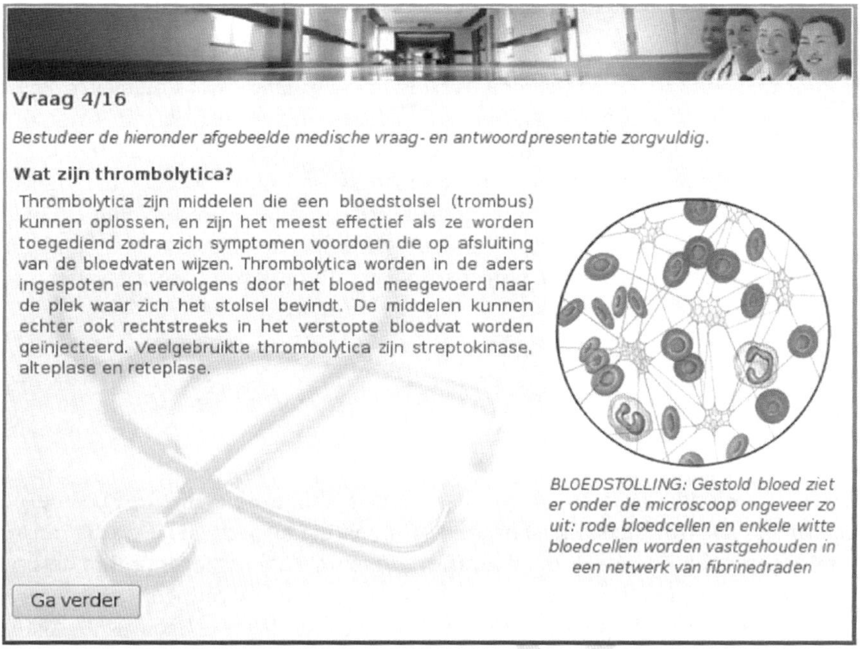

Fig. 6 Example of an answer presentation consisting of text and an automatically selected picture. The presentation answers the question "What are thrombolytics?" The text of the answer explains that thrombolytics are drugs used to dissolve blood clots. The picture depicts a schematic representation of clotted blood.

5.1 Participants

Seventy five people participated (44 female and 31 male). Their average age was 22 (SD = 7.11, min. = 18, max. = 55). Fifty six of them (75%) were students recruited from Tilburg University. None had participated in the previous two experiments.

5.2 Design

The experiment had a 16 (question) × 2 (short or long answer) × 2 (retrieval method: using caption or section) mixed factorial design, with the question as a within participants variable and the answer length and retrieval method as between participants variables. The dependent variables were the same as in Experiment 2, i.e., the participants' assessment of: (a) the clarity of the text, (b) the informativeness of the answer presentation, (c) the attractiveness of the answer presentation, (d) informativeness of the text-picture combination and (e) the attractiveness of the text-picture combination. The participants were randomly assigned to an experimental condition.

One of the goals of Experiment 3 was to compare the automatically illustrated answer presentations to the manually created answer presentations used in Experiment 2; therefore it reused the same design. Experiment 2 used manually selected pictures only, and relevance of the pictures was assumed. In contrast, some of the automatically selected pictures used in Experiment 3 were irrelevant, either because there was no appropriate picture in the database or simply because the algorithm failed to find one. However, choosing to use the same design for both evaluation experiments meant that in Experiment 3, the participants judged the informativeness of the text-picture combinations instead of directly assessing the relevance of the automatically selected pictures.

5.3 Stimuli

The study used the same set of 16 general medical questions that had been used in Experiment 2, with the same short and long textual answers. The textual answers were illustrated with automatically retrieved pictures using the algorithm described in Section 4. The pictures were retrieved from a repository of medical pictures that had been automatically extracted from two medical sources. Each of the pictures in the repository had two corresponding textual annotations: the first annotation represented the caption of the picture in the original document; and the second represented the paragraph (or section) in which the picture was found.

The pictures and their annotations were extracted from two medical sources intended for a general audience and written in Dutch, providing information about anatomy, processes, diseases, treatment and diagnosis. The first source, Merck Manual medisch handboek (Berkow et al, 2005), Merck in short, contains 188 schematic illustrations of anatomy and treatment, process schemas, plots and various types of diagrams. The other source, Winkler Prins medische encyclopedie (Fiedeldij Dop and Vermeent, 1974), WP in short, contains a variety of 421 pictures, including photographic pictures, schema's and diagrams. These sources were selected because they cover the popular medical field and they are relatively structured - paragraph boundaries are marked in the text and all 609 pictures have captions. The pictures have a high information density; only a few pictures are decorative. Consequently, the pictures are relatively specific to their context, which complicates their reuse in a slightly different context.

For each of the textual answers, two answer presentations were generated. For one of the presentations, the picture was retrieved using its caption as associated text, and for the other the picture was retrieved based on the smallest unit of surrounding text (paragraph or section) from the original document of the picture. Regardless of which of the texts was used for selecting the picture (caption or surrounding text), the caption was always presented together with the picture in the answer presentation. However, in order to prevent excessive caption lengths, captions were truncated to their first sentence during presentation generation (the remaining sentences were used for retrieval but not in the presentation). If the

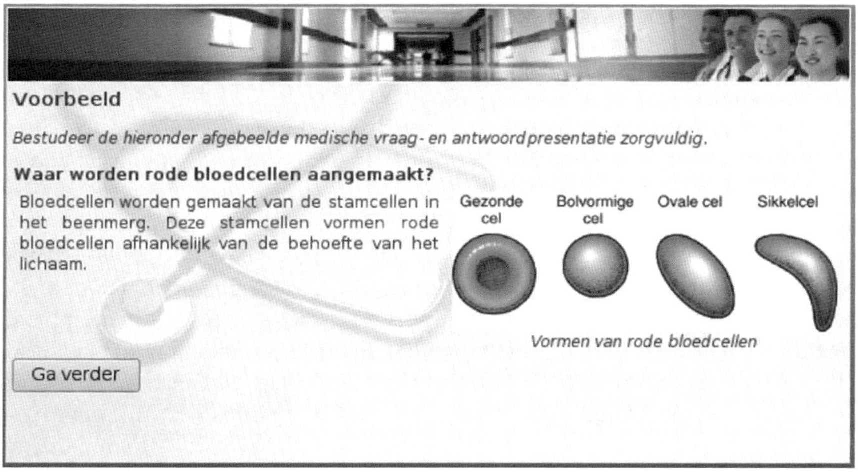

Fig. 7 Example of a picture which is related to, but not fully relevant to the answer text. The presentation answers the question "Where are red blood cells generated?" The text explains that red blood cells are generated from stem cells in the bone marrow. Rather than illustrating this, however, the picture shows various deformations of red blood cells.

surrounding text (section in short) was used for picture selection, this text was not included in the answer presentation. The corpus did not contain an appropriate picture for all answers, which forced the illustration system to select less appropriate pictures for some of the presentations. In some cases the selected picture was plainly irrelevant, but in some other cases, the picture was related to the text but had a different perspective. For instance, the picture in Figure 7 addresses the deformation of red blood cells rather than their generation. In the estimation (not formally validated) around 30% of the automatically selected pictures used in Experiment 3 were irrelevant, in the sense that they had absolutely no connection with the text of the answer. For example, a picture of egg and sperm cells was selected to illustrate an answer about RSI. The other pictures were either fully relevant, such as the picture in Figure 6, or somewhat relevant, such as the picture in Figure 7.

5.4 Procedure

The procedure was identical to the procedure of Experiment 2; see Section 3.4.

5.5 Data Processing

The results of the assessments were normalised to be in the range [0 ... 1]. A rating n between one and seven (inclusive) was normalized as $(n - 1)$. For processing the results, the following non-standard method was used. For each condition and each medical question and assessment question, the average assessment was calculated. For pair-wise significance testing of differences between two experimental conditions for a particular assessment question, the percentage of answer presentations was measured for which the rating of one condition was higher than that of another. A condition that consistently received higher than average ratings for each medical question got a score of 100%; consequently, the other condition got a relative score of 0%. Significance was tested by means of a 106-fold approximate randomisation. A difference is considered significant if the null hypothesis (that the sets are not different) can be rejected at a certainty greater than 95% ($p < .05$), unless stated otherwise.

The reasons for using the mutual rank instead of the average judgement were that the standard deviation of ratings of answers to some medical questions was higher than the standard deviation for answers to other medical questions. As a result, some medical questions affected the average rating more than others. This made it less likely to find significant differences in the average rating. Using the mutual rank avoided this problem.

5.6 Results

Caption versus Section

Figure 8(a) shows that the level of clarity of the textual component of the answer was judged similar. No significant differences between the conditions were found. Figure 8(b) indicates that for the informativeness of the presentation, long answers were rated significantly more informative than short answers. However, for long answers, the combination of picture and text (Figure 8(d)) was judged less informative. This difference was biggest for section-selected pictures, although not significant. Figure 8(c) and (e) show that the presentation as well as the picture-text combination were rated significantly more attractive if the pictures were selected by their captions than when the surrounding section was used for picture selection. No differences were found between short and long textual answers in the attractiveness of the presentation nor the picture-text combination.

Automatically versus Manually Selected Pictures

The results of two experiments are comparable only if the group of participants in one experiment is similar to the participants in the other experiment. In both Experiments 2 and 3, students and non-students took part, and their answers to

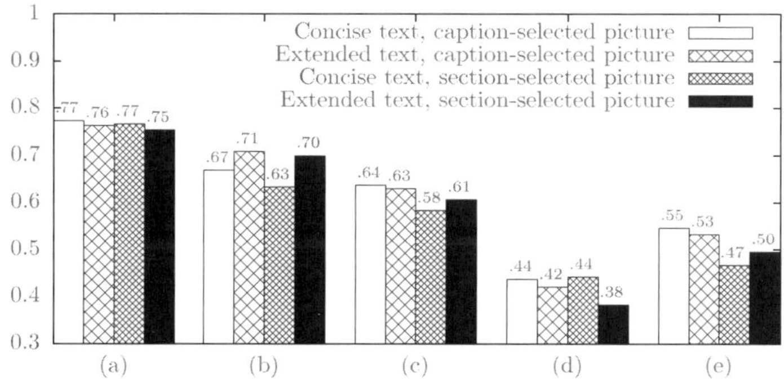

Fig. 8 Average assessments of (a) textual clarity; (b) informativeness of the answer presentation; (c) attractiveness of the answer presentation; (d) informativeness of the text-picture combination, and (e) attractiveness of the text-picture combination.

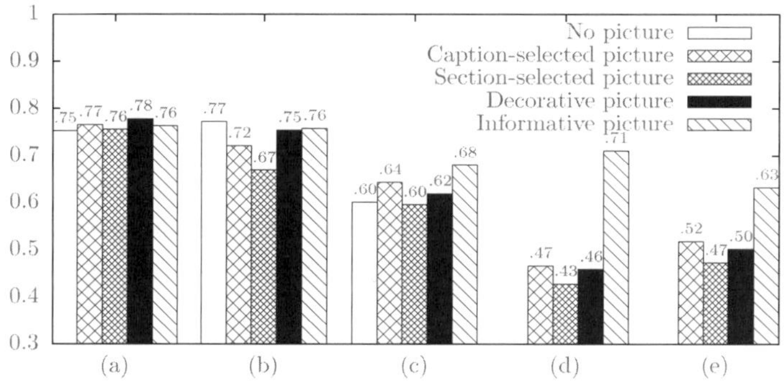

Fig. 9 Average assessments of (a) textual clarity; (b) informativeness of the answer presentation; (c) attractiveness of the answer presentation; (d) informativeness of the text-picture combination, and (e) attractiveness of the text-picture combination. For comparability, these results include only registered students from Tilburg University. Therefore, the actual values may differ slightly from Figure 8.

some of the assessment questions were significantly different. Therefore, to enable comparison of the two experiments, the group of non-students was filtered out, in order to ensure that the experimental conditions were the only variables over both experiments. In total, 98 people (70 female, 28 male) who participated in either Experiment 2 or 3 were registered students. Forty-two of them contributed to Experiment 2, and 56 contributed to Experiment 3. The average assessments of the 98 participants are shown in Figure 9.

These results combine the 16 short and the 16 long answer presentations, comprising 32 data points for each condition and assessment question. They include the unimodal condition from Experiment 2, which was not discussed in Section 3.

For informativeness of the answer presentation, no significant differences were found between answer presentations with a caption-selected picture and answer presentations with a manually selected informative picture. However, answer presentations with a section-selected picture were rated as significantly less informative than answer presentations with a manually selected informative picture, a decorative picture, or no picture at all. For attractiveness of the answer presentation, no significant differences were found between answer presentations with an automatically selected picture (either caption- or section-based), a manually selected decorative picture, or no picture at all. No significant effect was measured from the presence of (different types of) images on the user's perception of the clarity of the text.

The informativeness as well as the attractiveness of the text-picture combination was not significantly different between answers with an automatically selected picture (either caption- or section-based) or a manually selected decorative picture. However, the informativeness of the text-picture combination was rated significantly higher for answer presentations with a manually selected informative picture in relation to the answer presentations with an automatically selected picture or a manually selected decorative picture. Participants also found manual informative pictures more attractive than any other category in combination with the text.

Average ratings of automatic presentations may have been negatively affected by inconsistent performance of the picture selection algorithm. If the relevance of automatic pictures is less consistent than that of manual pictures, this should be reflected in the variability of the results. Indeed it was found that for automatic pictures, participants showed greater variability than for manual pictures in their assessments of textual clarity, informativeness and attractiveness of the answer presentation.

Cosine Similarity as Indicator of Picture Relevance

The selection criterion for automatic pictures was the cosine similarity of the textual component of the answer and the text associated with the picture (a caption or a section depending on the condition). The picture with the highest cosine similarity was selected. Because cosine similarity is used as a measure of relevance, this value can be interpreted as a confidence value, i.e. how confident the system is that the selected picture is actually relevant. In the IMIX system, in which this picture selection method is implemented, the answer is presenting text-only if no picture has a confidence (cosine similarity) above a certain configurable threshold. Table 4 shows the average of the cosine similarity values of the pictures selected for the answers in this experiment.

But what is the meaning of cosine similarity as a confidence value? Cosine similarity can be used to predict the relevance of the picture if there is a correlation between the cosine similarity and the experimental participants' judgements of a presentation. Figure 10 shows the correlation of the confidence (cosine similarity) value and the participant judgements. A value of 1 (or -1) indicates a perfectly increasing (or decreasing) linear correlation. This correlation was greatest for the participant judgements of the informativeness of the text-picture combination (.51 and .44 with short and long answer texts respectively). This is an encouraging result,

Table 4 Statistics of the cosine similarity of the textual component of the answer and the text passage used for indexing the selected picture.

Condition	Average	Standard deviation	Range
Brief text; caption-selected picture	0.190	(0.00788)	[0.0687,0.347]
Extended text; caption-selected picture	0.188	(0.00631)	[0.0786,0.397]
Brief text; section-selected picture	0.133	(0.00501)	[0.0295,0.311]
Extended text; section-selected picture	0.162	(0.00654)	[0.0373,0.319]

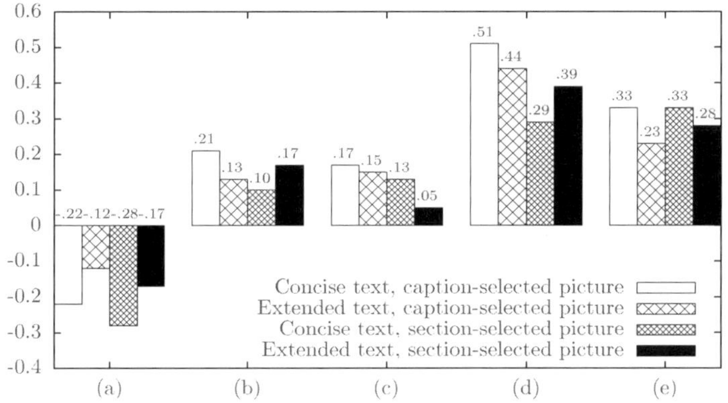

Fig. 10 Pearson correlation coefficient between the confidence (cosine similarity) of picture selection and the assessments of (a) textual clarity; (b) informativeness of the answer presentation; (c) attractiveness of the answer presentation; (d) informativeness of the text-picture combination, and (e) attractiveness of the text-picture combination.

given that this aspect seems to correspond most closely to picture relevance. With respect to attractiveness, the correlation with confidence was significantly greater for short answers than for long answers. There was only a slight difference in correlation between attractiveness and confidence for different picture selection methods.

Remarkably, participants perceived the textual component of answers as less clear when the confidence value of the picture was greater. This puzzling result suggests that relevant pictures negatively affect the clarity of the textual answer rather than enhance it. A possible explanation is that any mismatch between picture and text may be more confusing when text and picture seem closely related than when the picture obviously does not fit the text, in which case it can be easily ignored and does not influence the interpretation of the text.

5.7 Conclusion

The results of the evaluation experiment indicate that the caption-based picture selection method results in more informative and attractive presentations than the section-based method, although the difference in informativeness was not significant. Furthermore, caption-based picture selection shows a greater correlation between confidence and informativeness, which indicates that the confidence value better predicts the informativeness of the picture. Compared to manually created answer presentations, it was found that answer presentations with an automatically selected picture were largely rated at the same level as presentations with a manually selected decorative picture or even no picture at all. This is not entirely surprising. In Experiment 3, the manually selected pictures used in Experiment 2 were used as a gold standard for decorative and informative pictures respectively. However, in practice, it is unlikely that this gold standard could be achieved with the set of 609 pictures from the medical corpus, because the picture sources used by the participants in Experiment 1 (which formed the basis for the answer presentations in Experiment 2) were unrestricted and thus offered far more opportunities to find a suitable illustration for a given answer text.

Finally, an investigation of the relation between system confidence and the experimental results revealed a negative correlation between textual clarity and the predicted relevance of the selected illustration.

6 General Discussion

This chapter described three experiments investigating which (combinations of) presentation modes are most suitable for the answers of a medical QA system. In Experiment 1, we were interested in the spontaneous production of multimodal answers to medical questions. The results showed that people used pictures more frequently when producing long answers. Informative pictures were more frequently used in short answers, while representational pictures were most frequent in long answers. It is likely that when the answer does not contain much text, a picture will contain additional information with regard to the text. When the answer contains a lot of text, it is likely that a picture adds less information to it (i.e. it visually represents the information already present in text). Short answers contained more decorative pictures than long answers, possibly because the lack of room for discussing pictured information in short answers led the participants to add simple illustrations, requiring no textual explanation, more often than when creating presentations with long answers.

Also, people used decorative pictures more frequently in the answers to *what* questions. Informative pictures, on the other hand, occurred most often in the answers to *how* questions. Possibly, in textual answers to *what* questions the picture represented an element of the question. Pictures in the answers to *how* questions were often used to explain the steps within the procedure and therefore added information to the textual answer.

In Experiment 2, we concentrated on how people evaluate different multimodal (text and a picture) answer presentations on their informativeness and attractiveness. The results showed that answers with an informative picture were evaluated as more informative than those with a decorative picture. Moreover, *how* answers with informative pictures were evaluated as more informative than *what* answers with informative pictures. An explanation for this result could be that medical procedures – as they occurred in this experiment – lend themselves well to being visualised, as they have a temporal and spatial character. Definitions on the other hand often contain abstract concepts which are less easily visualised.

Another interesting result is that while short answers with an informative picture were evaluated as most informative, long answers with an informative picture were evaluated as most attractive. The information load of the textual answers might explain these results. Short and long textual answers differ in their information density, i.e. short answers contain less information than long ones. Therefore, an informative picture has more added value for short answers than for long answers, increasing the perceived informativeness of the short answer presentations. On the other hand, an informative picture adds relatively less information to a long textual answer and therefore primarily serves to enhance the attractiveness of the presentation.

In Experiment 3, we conducted a user evaluation in which two versions of the automatic picture retrieval method were compared: caption-selected illustrations versus section-selected illustrations. The caption-based picture selection method resulted in more informative and attractive answers than the section-based method, although the difference in informativeness was not significant. Furthermore, caption-based picture selection showed a greater correlation between confidence and informativeness, which indicates that the confidence value better predicts the informativeness of the picture. A system could use this to respond by not offering any picture if no relevant picture is available (as done in the IMIX system). All in all, the caption-based picture selection method offers more promising results than the section-based selection method.

When compared to manually created answer presentations, it was found that answer presentations with an automatically selected picture were largely rated at the same level as presentations with a manually selected decorative picture (which did not add any information to the answer) or even with no picture at all. This may be partially explained by the design of the experiment, where the visual element of the answer presentations was not needed to answer the question, since the textual element contained all the required information. Also, the results were undoubtedly influenced by the fact that the picture corpus did not contain appropriate pictures for all answers, in which case the algorithm had no choice but to select an irrelevant picture.

An investigation into the relation between system confidence and the experimental results revealed an intriguing negative correlation between textual clarity and the predicted relevance of the selected illustration. Apparently, seeing an answer text in combination with a picture that is related to it, but not fully attuned to it, may be confusing for the user. Problems like these might be solved by the development of post-processing methods to adapt the textual and visual components of the answer presentation to each other, so that they form a more coherent whole.

References

Arens Y, Hovy E, Vossers M (1993) On the knowledge underlying multimedia presentations. In: Maybury M (ed) Intelligent Multimedia Interfaces, AAAI Press

Berkow R, Beers MH, Fletcher AJ (eds) (2005) Merck manual medisch handboek, 2nd edn. Bohn Stafleu van Loghum, Houten, the Netherlands

Bernsen N (1994) Foundations of multimodal representations. a taxonomy of representational presentation mode. Interacting with Computers 6(4):347–371

Bosma W (2005) Image retrieval supports multimedia authoring. In: Zudilova-Seinstra E, Adriaansen T (eds) Linguistic Engineering meets Cognitive Engineering in Multimodal Systems, ITC-irst, Trento, Italy, ICMI Workshop, pp 89–94

Bosma W, Theune M, van Hooijdonk C, Krahmer E, Maes A (2008) Illustrating answers: an evaluation of automatically retrieved illustrations of answers to medical questions. In: Proceedings of the AISB Symposium on Multimodal Output Generation (MOG 2008), pp 34–41

Bosma W, Marsi E, Krahmer E, Theune M (2010) Text-to-text generation for question answering. In: Bouma G, van den Bosch A (eds) Interactive Multi-modal Question Answering, Springer Verlag

Carney R, Levin J (2002) Pictorial illustrations still improve students' learning from text. Educational Psychology Review 14(1):5–26

van Deemter K, Power R (2003) High-level authoring of illustrated documents. Natural Language Engineering 2(9):101–126

Fiedeldij Dop P, Vermeent S (eds) (1974) Winkler Prins medische encyclopedie, 3rd edn. Spectrum

Heiser J, Phan D, Agrawala M, Tversky B, Hanrahan P (2004) Identification and validation of cognitive design principles for automated generation of assembly instructions. In: Proceedings of Advanced Visual Interfaces, pp 311–319

van Hooijdonk C, Krahmer E (2008) Information modalities for procedural instructions: the influence of text, static and dynamic visuals on learning and executing rsi exercises. IEEE Transactions on Professional Communication 51(1):50–62

van Hooijdonk C, Krahmer E, Maes A, Theune M, Bosma W (2007a) Towards automatic generation of multimodal answers to medical questions: a cognitive engineering approach. In: Proceedings of the Workshop on Multimodal Output Generation (MOG 2007), pp 93–104 van Hooijdonk C, de Vos J, Krahmer E, Maes A, Theune M, Bosma W (2007b) On the role of visuals in multimodal answers to medical questions. In: Proceedings of the 2007 Conference of the IEEE Professional Communication Society, IEEE

de Jong FMG, Westerveld T, de Vries AP (2007) Multimedia search without visual analysis: the value of linguistic and contextual information. IEEE Transactions on Circuits and Systems for Video Technology 17(3):365–371

Maybury MT, Merlino AE (1997) Multimedia summaries of broadcast news. In: 1997 IASTED International Conference on Intelligent Information Systems, IEEE

Mayer RE, Moreno R (2002) Aids to computer-based multimedia learning. Learning & Instruction 12(1):107–119

Mayer RE (2005) The Cambridge handbook of multimedia learning. Cambridge University Press, Cambridge

Michas I, Berry D (2000) Learning a procedural task: effectiveness of multimedia presentations. Applied Cognitive Psychology 14(6):555–575

Nagao K, Ohira S, Yoneoka M (2002) Annotation-based multimedia summarization and translation. In: Proceedings of the 19th international conference on Computational linguistics, Association for Computational Linguistics, Morristown, NJ, USA, pp 1–7

Petrushin VA (2007) Introduction into Multimedia Data Mining and Knowledge Discovery, Springer London, pp 3–13

Porter M (1997) An algorithm for suffix stripping. In: Jones KS, Willet P (eds) Readings in Information Retrieval, Morgan Kaufmann, pp 313–316

Tversky B, Morrison J, Bétrancourt M (2002) Animation; can it facilitate? Int J Human-Computer Studies 57(4):247–262

Veenker T (2005) WWStim: A CGI script for presenting webbased questionnaires and experiments. Website: `http://www.let.uu.nl/~Theo.Veenker/personal/projects/wwstim/doc/en/`

Text-to-Text Generation for Question Answering

Wauter Bosma, Erwin Marsi, Emiel Krahmer and Mariët Theune

Abstract When answering questions, major challenges are (a) to carefully determine the content of the answer and (b) phrase it in a proper way. In IMIX, we focus on two text-to-text generation techniques to accomplish this: content selection and sentence fusion. Using content selection, we can extend answers to an arbitrary length, providing not just a direct answer but also related information so to better address the user's information need. In this process, we use a graph-based model to generate coherent answers. We then apply sentence fusion to combine partial answers from different sources into a single more complete answer, at the same time avoiding redundancy. The fusion process involves syntactic parsing, tree alignment and surface string generation.

1 Introduction

Answering specific types of trivia style (so-called 'factoid') questions is often taken as the core domain of *question answering* (QA) research. An example of such a question would be: *what is RSI?* But what is the correct answer to such a question? Ostensibly, this is a definition question, and a plausible answer is something like *RSI means Repetitive Strain Injury*. But will this answer the need for information of the person asking the question? In general, it seems that even if an unambiguous

Wauter Bosma
VU University Amsterdam, Amsterdam, The Netherlands, e-mail: w.bosma@let.vu.nl

Erwin Marsi
Norwegian University of Science and Technology, Trondheim, Norway, e-mail: emarsi@idi.ntnu.no

Emiel Krahmer
Tilburg University, Tilburg, The Netherlands, e-mail: e.j.krahmer@uvt.nl

Mariët Theune
University of Twente, Enschede, The Netherlands, e-mail: m.theune@ewi.utwente.nl

question is posed, users appreciate more information than a simple direct answer (Strzalkowski et al, 2000). Someone querying a system about RSI may be interested to know what the abbreviation stands for, but may also like to know what it actually *is*. Bates (1990) helps explain the findings of Strzalkowski et al by viewing an information search as a 'berry picking' process. Consulting an information system is only part of a user's attempt to fulfil an information need. It's not the end point, but just one step whose result may motivate a follow-up step. The 'factoid answer approach' fails to show leads to related information, which may trigger follow-up questions. Bakshi et al (2003) show that when answering questions, increasing the amount of text returned to users significantly reduces their number of queries, suggesting that users utilise related information from the text surrounding the answer.

In short, the raw response of a Question Answering system is often not a suitable answer. Text-to-text generation can help transforming a QA response into an appropriate answer. Generating an answer involves two core decisions to be made: (1) which information should be included, and (2) how that information should be presented. Decision (1) requires *content selection*, which is the process of finding the boundary between useful information and superfluous information. Decision (2) involves choosing an optimal formulation.

This chapter describes the efforts in text-to-text generation within the IMOGEN project. In particular, it describes two focus areas of research to improve the quality of the answer: (a) graph-based content selection to improve the answer in terms of usefulness, and (b) sentence fusion to improve the answer in terms of formulation. Sentence fusion is used to join together multiple sentences in order to eliminate overlapping parts, thereby reducing redundancy. The results of this work have been applied in the IMIX system. This system uses a Question Answering system to pinpoint fragments of text which are relevant to the information need expressed by the user. A content selection system then uses these fragments as entry points in the text to formulate a more complete answer. Sentence fusion is applied to manipulate the result in order to increase the fluency of the text.

For example, for the question '*what is RSI*', the content selection system may find several passages which may be used in the answer:

1A RSI means *repetitive strain injury*.

1B Repetitive Strain Injury is generally caused by a mixture of poor ergonomics, stress and poor posture.

1C Repetitive Strain Injury may occur for instance in people who regularly work with a display device.

The system could return an answer by just enumerating the above sentences, but this answer would contain some degree of redundancy, because the term *repetitive strain injury* occurs in each of the sentences. By using sentence fusion, the last two sentences can be fused, generating a more fluent answer with less redundancy:

> RSI means *repetitive strain injury*. Repetitive Strain Injury is generally caused by a mixture of poor ergonomics, stress and poor posture, and may occur for instance in people who regularly work with a display device.

This chapter presents the progress made in developing text-to-text generation techniques for Question Answering. The next two sections—Section 2 and 3—are dedicated to content selection and sentence fusion respectively. The chapter closes with some final thoughts and an outlook in Section 4.

2 Graph-based Content Selection

Not boring anyone with irrelevant details and, at the same time, not withholding the essentials. This is the essence of content selection. It can also be seen as an optimisation issue: more information takes the user more time to process, while less information may increase the number of interactions. The greatest efficiency is achieved when precisely that information is given, which is relevant to the user.

This section describes a framework for content selection which is based on the notion of *contextual salience* – all evidence of salience of a particular content unit is based on the salience of related content units (its context). The underlying data model is based on graph theory – a paradigm which is excellently equipped for representing relations between content units, thus modelling coherence and redundancy in text.

The model separates the actual content selection from the detection of coherence and redundancy relations. The relation detection process results in a graph, and the content selection algorithm uses that graph to select the sentences to include in the summary. This separation makes it possible to replace the relation detection algorithms or the content selection algorithm without changing anything else. It is also possible to combine several relation graphs and use the combined graph for content selection.

As such, the graph-based model does not prescribe the nature of the relations or algorithms used to find the relations: the relations may represent coherence relations, co-reference relations, lexical chains, cosine similarity, or any other paradigm that may model (semantic) relations in text and that is compatible with the graph-based representation. Each of those methods may be individually implemented and evaluated, possibly in combination with other methods. The model was evaluated using sentences as content units and cosine similarity as a feature to find relations between sentences.

2.1 Related Work

Content selection – choosing what to include and what not to include in a summary – is useful in answer presentation; it is also a subtask of summarisation. A typical text summarisation system transforms a source document (or a set of documents) into a more concise document by: (1) splitting up the source into individual *content units*, (2) selecting the most salient content units, and (3) composing a summary of those content units. The system may exploit contextual information such as a user question. Because of the availability of resources and tools in automatic summarisation, algorithms are evaluated for content selection in the context of a summarisation system.

Content units may vary in granularity from paragraphs to phrases. Often, sentences are chosen as content units because they are reasonably fine-tuned information units and, at the same time, the possibility of ungrammatical results can be avoided (which is a major challenge in phrase-based summarisation).

Rather than viewing a summary as a single text, summarisation systems typically perform content selection by determining the relevance of each passage independently, and then composing a summary of the top ranking passages. Classical features for scoring sentences include the presence of cue phrases, term frequency, stop word lists, etc. (Edmundson, 1969; Luhn, 1958). Assessing the relevance of each sentence individually, these systems neglect the internal structure of the summary, despite insights into discourse organisation which claim that meaning is tightly related to discourse organisation (e.g. Mann and Thompson, 1988). The meaning in a text is not merely the sum of the meaning into its passages, but a passage should be interpreted in the context shaped by other passages. For example, given the two passages below, the second passage would have little meaning if the context provided by the first were omitted. Hence, a generic summarisation system should include the second sentence in a summary only if the first is also included.

2A A commercial airliner crashed in northwestern Iran on Wednesday.

2B All 168 people on board were killed.

If content selection is to result in an answer to a user query or question, this makes the need for dealing with coherence even more pressing. While a generic summary (i.e., a summary which is generated without user input) should be internally coherent, a query-based summary should also be coherent with respect to the query. Similarly, the task of multi-document summarisation (if the answer is drawn from multiple documents) introduces the need to deal with redundancy, as a summary should not mention the same thing twice.

A number of ad-hoc solutions to dealing with redundancy and coherence emerged in response to the challenges of multi-document and query-based summarisation. For instance, Carbonell and Goldstein (1998) introduced the concept of *marginal relevance* to handle redundancy. They build up a summary by adding sentences one by one, with a bias toward sentences which contain new information in relation to already-selected sentences.

Barzilay and Elhadad (1997) modelled coherence by dividing the source into topics by identifying *lexical chains*. They composed summaries of one sentence from each of the strongest topics, so as to maximise coverage. The summarisation system of Blair-Goldensohn and McKeown (2006) prioritises sentences in the summary which have a coherence relation to another summary sentence. Each of these answers to the problem of coherence represents a minor change to an existing summarisation system, rather than an integral model of coherence. Other summarisation systems (e.g. Marcu, 1999; Wolf and Gibson, 2005) integrate a more

Title:	former President Carter's international activities
Query:	Describe former President Carter's international efforts including activities of the Carter Center.

Fig. 1 A DUC 2006 topic (D0650E).

sophisticated model of coherence in the content selection process, but they require high level semantic annotation which can (for now) only be achieved manually.

2.2 Task Definition

The Document Understanding Conference (DUC) 2006 data set was used for training, and the DUC 2005 data set for testing.[1] This is possible because both data sets are similar. The task at hand is to automatically generate a summary of a maximum of 250 words, given a *topic*. A topic consists of a title, a query, and a set of source documents. The summary should answer the query using the source documents. An example of a topic is given in Figure 1. The DUC 2006 document set consists of 50 topics with 25 source documents each. The DUC 2005 document set consists of 50 topics with 25–50 source documents each (approximately 32 on average).

The summarisation task is given to professional human summarisers as well as to automatic summarisation systems. The human summaries are used as *reference summaries* for evaluating *candidate summaries* (i.e., generated summaries). Each DUC 2005 topic has six corresponding reference summaries; each DUC 2006 topic has four. Rouge-2 recall (i.e. bigram recall with respect to reference summaries) and Rouge-SU4 recall (skip bigram recall) was used as performance metrics for evaluation (Lin, 2004), because these metrics were also used (with the same configuration) at DUC 2005 and DUC 2006. Although Rouge metrics provide only a partial evaluation of a summarisation system, they are reproducible and repeatable for different system configurations in an objective manner, since they require no manual intervention.

To measure if one summarisation algorithm performs better (or worse) than another with a particular metric, the number of topics is counted for which it outperformed the other, and vice versa. Then, an approximate randomisation test is run to measure statistical significance (Noreen, 1989).

[1] Available from `http://duc.nist.gov`

2.3 A Framework for Summarisation

The goal is to investigate new methods for content selection. Nonetheless, the evaluation methods used are designed to measure the quality of abstracts, and require a full summarisation system. The summarisation system is briefly characterised, and then the content selection components are focused upon. The summarisation system consists of the following components.

Segmentation. The source documents as well as the query are segmented into sentences. The document name, the paragraph number and sentence number are associated with each sentence as meta-information. The document name can later be used to detect whether sentences are from the same document, or whether they are query sentences. Paragraph boundaries are derived from annotations provided with the source documents. The segmenter also attempts to remove meta data from the text, such as the date and location of publication. These meta data are not part of the running text and may introduce noise in the summary.

Feature extraction. The source text and the query are processed and converted to a feature graph to prepare for content selection. Multiple modules may be used in parallel so that multiple graphs are generated. This may include coherence analysis, measuring redundancy, etc. The generated graphs are integrated into a combined graph, as described later in this chapter.

Salience estimation. A salience value is derived for each sentence from the (possibly combined) feature graph.

Presentation. A summary is created using the most salient content units, up to the word limit of 250 words. If adding the next-salient sentence would cause the word limit to be exceeded, no more sentences are added. Where possible, the linear ordering of the sentences in the source text is retained. If the summary contains sentences from multiple source documents, sentences from the document containing the largest number of sentences are presented first. Although the ordering of the sentences may be important for readability, it has little effect on the Rouge scores.

The next section compares different methods for *feature extraction* and *salience estimation*. Across these experiments, the components of *segmentation* and *presentation* remain unchanged.

2.4 Query-based Summarisation

Below, a number of systems is described by the summarisation framework they adopt, starting with a rudimentary summarisation system, and adding features to build increasingly sophisticated systems. The modular summarisation framework allows for the flexibility to add feature graphs or replace the salience estimation algorithm.

Wherever parameter optimisation is used, the DUC 2006 data set is used for this purpose. The results are then validated with the DUC 2005 data set.

2.4.1 Query-relevance

A simple form of query-based summarisation is to determine sentence salience by measuring its cosine similarity with the query. The sentences most similar to the query are presented as a summary. This constitutes a competitive baseline system for query-based summarisation. The graph used for salience estimation is the graph in which each candidate sentence is related to each query sentence, and the strength of this relation is the cosine similarity of the two sentences. The sentences closest to a query sentence are then included in the summary. The cosine similarity graph is generated in three steps:

1. the words of all sentences are stemmed using Porter's stemmer (Porter, 2001);
2. the inverse document frequency (IDF) is calculated for each word;
3. the cosine similarity of each candidate sentence and each query sentence is calculated using the $tf \cdot idf$ weighting scheme.

Stemming is a way to normalise morphological variation. For example, the words *cause* and *causes* are different words but refer to the same concept. Stemming allows both forms to be treated in the same way.

The inverse document frequency is used to assign a greater weight to words which occur in fewer sentences. Rare words typically characterise the sentence they appear in to a greater extent than frequent words. IDF is a way to account for this (Spärck Jones, 1972).

IDF calculation requires statistics from the language to determine the frequency of words. For calculating the IDF values of the words in source documents, the statistics of the set of all source documents in the topic are used. It is inappropriate to use the same statistics to weight the query words, because there is a mismatch between the language use in the query and in the source documents. For instance, queries frequently use phrases such as 'Discuss ...' or 'Describe ...'. These words have a low frequency in the source documents, and thus would be assigned a high IDF value, although they are hardly descriptive if they appear in the query. On the other hand, a single query is too short to draw useful statistics from. Therefore, the IDF values for query terms are calculated from the set of sentences from all DUC 2006 queries, while the IDF values for source documents are calculated from the set of source document sentences specific for the topic.

The query-relevance graph (called δ_q) is defined by a function determining the strength of the relation between two sentences:

$$\begin{aligned} \delta_q(i,j) &= cosim(i,j) && \text{, if } i \in Q;\ j \in S \\ \delta_q(i,j) &= 0 && \text{, otherwise} \end{aligned} \tag{1}$$

where $\delta_q(i,j)$ is the strength of the relation between sentences i and j; Q is the set of query sentences; S is the set of candidate sentences; $cosim(i,j)$ is the cosine similarity of sentences i and j. The strength of a relation is a value in the range of 0 (no relation) to 1 (a strong relation).

The query-relevance $R_q(j)$ of a sentence j is then calculated by taking the greatest strength of any relation of j to a query sentence:

$$R_q(j) = \max_{q \in Q} (\delta_q(q,j)) \tag{2}$$

where $R_q(j)$ is the salience of sentence j; Q is the set of query sentences.

A summary is then generated from the most salient sentences.

2.4.2 Contextual Relevance

The *cohesion graph* (δ_c) is added as a feature graph for calculating contextual relevance. This graph is constructed in the same way as the query-relevance graph, except that it relates candidate sentences of the same document, rather than query sentences and candidate sentences.

The graphs δ_q and δ_c are integrated into a single multigraph Δ_{q+c}. A multigraph-graph is a graph that can have two edges between the same two sentences, expressing simultaneous relations. As a result, not a single relation but a set of relations hold between two sentences, and each relation may have a different strength between 0 and 1. The integrated graph is expressed as follows.

$$\Delta_{q+c}(i,j) = \{w_q\delta_q(i,j), w_c\delta_c(i,j)\} \tag{3}$$

where $\Delta_{q+c}(i,j)$ is a set of values, each representing the strength of an edge from i to j in the multigraph Δ_{q+c}. The values of $w_q, w_c \in [0..1]$ are weighting factors. The smaller the w_q and the greater the w_c, the greater the relative importance of indirect evidence of relevance. A greater w_q (relative to w_c) results in selecting more sentences which can be directly related to the query. A greater w_c represents a greater bias toward sentences which are relevant indirectly, and which may increase the coherence of the summary as a whole.

The salience estimation algorithm calculates the salience of each sentence, given a graph of relations between sentences. A relation from sentence X to sentence Y increases the relevance (and therefore the salience) of Y if X is relevant. This immediately poses a problem if X is a candidate sentence, because the relevance of Y depends on the relevance of X, which itself is not yet calculated. Literature provides several solutions (Mani and Bloedorn, 1997; Erkan and Radev, 2004), which have in common the fact that they iteratively recalculate the salience of a sentence in a graph from the salience of neighbouring sentences. Following this process, salience is calculated as follows.

1. Initiate the salience of all candidate sentences (source document sentences) at 0. The salience of query sentences is initiated at 1.

2. Recalculate the salience of each candidate sentence, using the feature graphs and the salience of neighbouring (i.e. related) sentences. Salient sentences increase the salience of their neighbours.
3. Repeat step 2 until the sum of changes in salience in the last iteration falls below a certain (pre-defined) threshold.

Two salience estimation algorithms are used. These are *normalised centrality* and *probabilistic relevance*. They differ in how they recalculate relevance (step 2).

The first, based on Erkan and Radev (2004), recalculates the salience by dividing the salience of each sentence among its neighbouring sentences. Because the sum of salience values of all sentences remains approximately constant (amounts of salience are just "passed on" from one sentence to the next), this is called *normalised centrality*.

The *probabilistic relevance* algorithm regards the feature graph as a probabilistic semantic network. The salience of a sentence represents the probability that the sentence is relevant, and a relation from sentence X to Y is the probability that Y is relevant, given X is relevant.

Normalised centrality. The result of the normalised centrality process is a *centrality* value for each sentence, which represents the salience of the sentence. The basic idea behind this algorithm is that at each iteration, the centrality value of each sentence is distributed among its related sentences. For instance, if a sentence has three outgoing relations, the centrality value is divided amongst these three sentences. Conversely, the new centrality value of a sentence at the next iteration is calculated from the sum of "centrality" received from sentences to which it has an incoming relation.

The symbol μ_j was used to indicate the centrality of sentence j. The value of $\mu_j(t)$ is the centrality of sentence j at iteration $t \geq 0$, which is calculated as follows:

$$\begin{aligned}
\mu_j(t) &= 1 && \text{, if } j \in Q \\
\mu_j(0) &= 0 && \text{, if } j \in S \\
\mu_j(t+1) &= \frac{d}{\|D\|} + (1-d)\sum_{i \in D} x(i,j) && \text{, if } j \in S \\
x(i,j) &= \sum_{r \in \Delta_{q+c}(i,j)} r \cdot \mu_i(t) \cdot degree(i)^{-1}
\end{aligned} \tag{4}$$

where $D = Q \cup S$; and $\Delta_{q+c}(i,j)$ is the set of edges between i and j in the relevance graph.

By definition, query sentences have a centrality of 1. In the first iteration, the centrality of all other sentences is initialised to 0. At each iteration, the normalised centrality of each sentence is distributed to its neighbour sentences (the sentences related to the sentence). The constant d is a small value, which is required in order to guarantee that the algorithm converges under all circumstances, by giving each sentence a small a priori non-zero centrality.[2] The degree of a sentence i in the graph is measured as the sum of the weights of the outgoing edges:

$$degree(i) = \sum_{k \in D} \sum_{(r \in \Delta_{q+c}(i,k))} r \tag{5}$$

A characteristic of this algorithm is that across iterations, the sum of the centrality of all sentences stays approximately the same: centrality is treated as a kind of commodity which may be transferred from one sentence to the next, but no centrality is created or lost. The only exceptions to this rule are the query sentences (which always have a centrality of one, and therefore may "create" centrality) and the constant d, which causes a small centrality value to be assigned to each sentence.

The result is a centrality (i.e., salience) value μ between 0 and 1 associated with each passage. The content units with the highest salience values are selected for inclusion in the summary. In this configuration, normalisation cancels out the effect of graph weighting: changing the graph weights w_q and w_c (cf. equation 3) does not affect the summaries in any way because the relevance distribution is normalised and the sets of sentences with outgoing edges in δ_q and δ_c are disjunct.

Probabilistic relevance. In the probabilistic approach, relations between sentences are interpreted as probabilities. The strength of a relation is the probability that the target sentence is relevant, provided that the origin sentence of the relation is relevant. In this algorithm, the salience calculation of a sentence at each iteration depends only on the salience values of its related sentences, and the strengths of these relations. This is unlike normalised centrality, where a sentence has to "compete" for a related sentence's centrality, since centrality is distributed among related sentences proportional to the relation strengths.

The query sentences are relevant by definition (although they are not actually included in the summary):

$$v_j(t) = 1 \qquad , \text{if } j \in Q$$

If a candidate sentence has only one incoming relation, its probabilistic relevance in the next iteration is the strength of the relation multiplied by the relevance of the origin sentence of the relation:

$$\begin{aligned} v_j(0) &= 0 && , \text{if } j \in S \\ v_j(t+1) &= r \cdot v_i(t) \cdot y && , \text{if } j \in S; t \geq 0 \end{aligned} \tag{6}$$

where $v_j(t)$ is the probabilistic relevance value of sentence j at iteration t, and r is the strength of the relation. The value of y is the *decay value*, a global constant in the range $\langle 0..1 \rangle$. The constant y has a function similar to the constant d in normalised centrality: it is necessary to ensure that the relevance value converges.

[2] Throughout this section, the value of 0.15 is used, as suggested in Erkan and Radev (2004), but the actual value of d has no effect on the final centrality ranking as long as it is non-zero.

If a sentence has more incoming relations, its relevance value depends on the relevance of the origin sentences of the relations. If $P(j \mid i)$ denotes the (conditional) probability that j is relevant, provided that i is relevant, then $P(\neg j \mid i) = 1 - P(j \mid i)$ is the probability that i is *not* relevant, provided that i is relevant. If all relation strengths are read as the conditional probability that the target sentence is *not* relevant (*inverse probability*), it is possible to combine multiple probabilities by multiplying their inverse probabilities, and then again taking the inverse of the result to get the combined probability:

$$P(i) = 1 - \prod_{j}(1 - P(i \mid j)) \tag{7}$$

Combining eq. 6 and 7, the relevance of a sentence j is calculated at each iteration as:

$$v_j(t+1) = 1 - \prod_{i \in Q \cup S} \prod_{r \in \Delta_{q+c}(i,j)} (1 - r \cdot v_i(t) \cdot y)$$

The relation set $\Delta_{q+c}(i,j)$ is the result of the union of the graphs δ_q and δ_c, and their weights w_q and w_c. The optimal weight values are estimated by measuring Rouge-2 performance for different weight values. First, w_q is incremented in steps of 0.1 from 0 to 1 with $w_c = 1$, and then w_c is incremented in steps of 0.1 from 0 to 1 with $w_q = 1$. The optimal weight settings are $w_q = 1$; $w_c = 0.1$.

2.4.3 Redundancy-aware Summarisation

One of the assumptions often made implicitly in the design of single-document summarisation systems, is that the source document does not contain redundancy. Consequently, there is no risk of including a sentence in the summary which does not contain any new information. This changes when a summary is generated from multiple source documents, where non-redundancy of sentences from different documents cannot be taken for granted. The content selection procedures outlined previously concentrate entirely on relevance, not on redundancy. However, in multi-document summarisation, presented content should be relevant to the query and novel with respect to what is already mentioned in the summary. In other words, salience comprises both relevance and novelty.

To accommodate representing novelty, the model extends with a redundancy feature graph Υ which is used in addition to the previously mentioned relevance feature graph Δ. Similarly to relevance, redundancy relations have a strength in the range $[0..1]$. The strength of a redundancy relation between two sentences expresses the likelihood that a sentence is redundant, given the fact that another sentence is relevant.

Although redundancy comes into play only during sentence selection, this way of modelling redundancy makes it possible to generate the redundancy graph in parallel with the relevance graph generation. That is, *before* the actual sentence selection, or even before the query is formulated. The actual novelty calculation still has to take place during sentence selection, but this is just a simple calculation while the redundancy graph may be the result of sophisticated processing algorithms. The redundancy graph generation algorithms do not need specific knowledge of summarisation and can focus on their isolated task, which is to calculate the probability that a sentence i is redundant, provided that a sentence j is relevant, for each pair of sentences i, j.

The redundancy of sentence j, given sentence i, is defined by $\delta_r(i,j)$. The form of the redundancy graph is identical to that of the relevance graph. The strengths of relations in the redundancy feature graph δ_r are defined as follows:

$$\begin{aligned} \delta_r(i,j) &= cosim(i,j) && \text{, if } i,j \in S;\ doc(i) \neq doc(j) \\ \delta_r(i,j) &= 0 && \text{, otherwise} \end{aligned} \tag{8}$$

The redundancy-aware summarisation system uses a set of redundancy feature graphs Υ for determining salience of sentences, in addition to the relevance feature graphs Δ:

$$\begin{aligned} \Delta_{q+c+r}(i,j) &= \left\{ w_q \cdot \delta_q(i,j), w_c \cdot \delta_c(i,j), w_{r\Delta} \cdot \delta_r(i,j) \right\} \\ \Upsilon_r(i,j) &= \left\{ w_{r\Upsilon} \cdot \delta_r(i,j) \right\} \end{aligned} \tag{9}$$

where δ_q, δ_c and δ_r are the query-relevance graph, the cohesion graph, and the redundancy graph respectively. The set of relations between sentences i and j is represented by $\Delta_{q,c,r}(i,j)$ (relevance) and $\Upsilon_r(i,j)$ (redundancy). Since redundancy implies 'relatedness', a redundancy graph is regarded as a special case of a relevance graph. Therefore, δ_r is not only included in Υ_r but also in Δ_{q+c+r} (with weights $w_{r\Upsilon}$ and $w_{r\Delta}$ respectively).

The calculation of redundancy-adjusted salience was inspired by Carbonell and Goldstein (1998). First, the relevance of each sentence is calculated using Δ_{q+c+r}. Then, the novelty is calculated – novelty is the reciprocal of redundancy. If two sentences are redundant, this affects only the novelty of the least relevant of the two. The implication of this is that a redundancy relation may cause a sentence to be downranked, but only if it is redundant with respect to a higher ranking sentence (which was already preferred over the downranked sentence during sentence selection). The stronger the redundancy relation, the greater the reduction of novelty. Novelty is calculated as follows:

$$N(j) = \prod_{i \in F_j} \prod_{r \in \Upsilon_r(i,j)} (1 - r \cdot R(i)) \tag{10}$$

$$F_j = \{k : S \mid R(k) > R(j)\}$$

where $N(j)$ is a value in the range $[0..1]$, representing the novelty of sentence j; $\Upsilon_r(i,j)$ is a set of redundancy relations, expressing the redundancy of j given i; F_j is the set of content units more relevant than j. The function $R(i)$ denotes the relevance of sentence i, as previously calculated.

Now, the redundancy-adjusted salience can be calculated as the product of relevance and novelty:

$$\sigma_j = R(j) \cdot N(j) \quad (11)$$

where σ_j is the redundancy-adjusted salience of sentence j. The calculation of σ_j ensures that:

- if one content unit is selected, all content units redundant to that unit are less likely to be selected: if two content units are redundant with respect to each other, the salience of the least-relevant content unit is reduced;
- redundancy of a content unit does not prevent relevance to propagate: a redundant content unit may still be relevant.

The graph weights are determined by starting from the optimal values for w_q and w_c, as specified in Section 2.4.2. The remaining weights are determined by means of a similar procedure as in Section 2.4.2: first, $w_{r\Delta}$ is incremented in steps of 0.1 from 0 to 1 with $w_{r\Upsilon} = 0$, and then $w_{r\Upsilon}$ is incremented in steps of 0.1 from 0 to 1 without changing the other weights.

For the normalised centrality algorithm, the resulting optimal weight settings are $w_q = 1$; $w_c = 1$ and $w_{r\Upsilon} = 0$; $w_{r\Delta} = 1$. Increasing the value of $w_{r\Upsilon}$ has no effect on the quality of the summaries. For the probabilistic relevance algorithm, the resulting optimal weight settings are $w_q = 1$; $w_c = 0.1$; $w_{r\Delta} = 0.2$; $w_{r\Upsilon} = 1$.

2.5 Results

Table 1 Performance on DUC 2006 data: Rouge scores, and the system rank among 36 systems (between brackets) if it had participated in DUC 2006.

System	Rouge-2	Rouge-SU4
Query-relevance	.08180 (11)	.1384 (11)
Normalised centrality	.08195 (11)	.1362 (11)
Probabilistic relevance	.08884 (3)	.1432 (7)
Redundancy-aware normalised centrality	.09294 (2)	.1496 (2)
Redundancy-aware probabilistic redundancy	.09305 (2)	.1501 (2)
Best DUC 2006 submission	.09505 (-)	.1546 (-)

Table 1 shows the Rouge scores for each of the summarisation systems. Table 2 gives an overview of the differences in performance between the systems. The query-relevance system was beaten by every other system ($p < 0.01$) except the normalised centrality system. The (redundancy-aware) probabilistic relevance system outperformed the (redundancy-aware) normalised centrality system ($p < 0.05$). The introduction of redundancy significantly improved results in the normalised centrality system ($p < 0.05$), but not in the probabilistic relevance system. The

Table 2 Percentage of DUC 2006 topics (Rouge-2/Rouge-SU4) for which one system (rows) beat another (columns). Note that percentages do not add up to 100 if both systems receive the same score for at least one topic.

%	(a)	(b)	(c)	(d)	(e)
(a) Query-relevance	–	50/52	34[a]/28[a]	30[a]/28[a]	26[a]/26[a]
(b) Normalised centrality	46/48	–	34[a]/36[b]	38[b]/34[a]	30[a]/24[a]
(c) Probabilistic relevance	64[a]/70[a]	66[a]/62[b]	–	56/58	44/50
(d) Redundancy-aware normalised centrality	66[a]/66[a]	60[b]/62[a]	42/42	–	30[a]/30[a]
(e) Redundancy-aware probabilistic relevance	70[a]/72[a]	68[a]/72[a]	48/46	64[a]/68[a]	–

[a] Significant at $p < 0.01$.
[b] Significant at $p < 0.05$.
[c] Significant at $p < 0.1$.

redundancy-aware probabilistic relevance system is the only system which beats all systems except the (non redundancy-aware) probabilistic relevance system ($p < 0.01$).

2.6 Validating the Results

The previous section outlined a comparison of the different configurations of the summarisation framework. However, the way the graph weight configurations are determined implies that the weights are tailored to the DUC 2006 data set. As a result, there is a risk that the weights are overfitted to this particular set. In order to validate the results, the experiments were run on the DUC 2005 data set with the graph weight configurations determined in Section 2.4.

Table 3 shows the average Rouge-2 and Rouge-SU4 scores achieved with the DUC 2005 corpus. Table 4 shows an overview of the pair-wise significance tests. The redundancy-aware probabilistic relevance system significantly outperforms all other systems when Rouge-2 is used ($p < 0.1$), and all except the redundancy-aware normalised centrality system according to Rouge-SU4. This system would have ranked first (Rouge-2) or second (Rouge-SU4) if it had participated in DUC 2005. The overall picture confirms the preliminary results in Section 2.5 with respect to the differences between normalised centrality and the probabilistic relevance: the latter system outperformed the first (Rouge-2: $p < 0.01$; although no significant differences were measured using Rouge-SU4), and the redundancy-aware variant of the probabilistic relevance system also outperformed the redundancy-aware normalised centrality system ($p < 0.01$, Rouge-2 and Rouge-SU4). However, introducing redundancy did not generally improve the results. Only in the probabilistic relevance system was a significant Rouge-2 improvement found ($p < 0.1$) as a result of the added redundancy graphs.

An interesting observation is that the probabilistic relevance system has the lowest average scores, but still beats the normalised centrality system in terms of the number of summaries for which the Rouge-SU4 score was greater. Apparently,

Table 3 Performance on DUC 2005 data: Rouge scores, and the system rank among 32 systems (between brackets) if it had participated in DUC 2005.

System	Rouge-2	Rouge-SU4
Query-relevance	.07056 (3)	.1260 (5)
Normalised centrality	.07210 (2)	.1262 (5)
Probabilistic relevance	.06923 (5)	.1247 (6)
Redundancy-aware normalised centrality	.07230 (2)	.1291 (3)
Redundancy-aware probabilistic relevance	.07362 (1)	.1300 (2)
Best DUC 2005 submission	.07251 (-)	.1316 (-)

Table 4 Percentage of DUC 2005 topics (Rouge-2/Rouge-SU4) for which one system (rows) beat another (columns). Note that percentages do not add up to 100 if both systems receive the same score for at least one topic.

%	(a)	(b)	(c)	(d)	(e)
(a) Query-relevance	–	46/44	42/42	50/50	40[c]/40[c]
(b) Normalised centrality	52/54	–	50/34[a]	50/54	38[b]/34[a]
(c) Probabilistic relevance	54/58	50/66[a]	–	58[b]/64[a]	36[c]/42
(d) Redundancy-aware normalised centrality	44/44	46/44	38[b]/36[a]	–	30[a]/30[a]
(e) Redundancy-aware probabilistic relevance	58[c]/60[c]	60[b]/66[a]	54[c]/54	60[a]/70[a]	–

[a] Significant at $p < 0.01$.
[b] Significant at $p < 0.05$.
[c] Significant at $p < 0.1$.

the probabilistic relevance system usually (in 66% of the cases) beat the normalised centrality system, but there were a few cases in which the score of the probabilistic system was considerably lower. This raises the question, which system is better: the one which receives the greatest average score, or the one which most frequently produces the better summary?

Note also, that there is no guarantee that the combination of graph weights leading to the best performance has been found. Apart from the risk of overfitting, the number of possible graph weight combinations is infinite, and a greater number of graphs makes it more difficult to find the best combination of weights. A future extension would use machine learning methods such as genetic algorithms that are better suited to find the optimal solution. Despite this, the current system already achieved good results, using this limited optimisation.

3 Sentence Fusion

Question Answering systems conventionally follow the strategy of retrieving possible answers from a text collection, ranking these answers according to some criteria of relevance, and outputting the top-ranked answer. Many current QA systems even rely on various parallel answer-finding strategies, each of which may produce an n-best list of answers (e.g. Maybury, 2004). However, the underlying assumption that a single complete answer can be pinpointed in the text collection is questionable at best. Especially if the question is of the *open* type rather than the *factoid* type (where the answer typically is some named entity), relevant parts of the answer may be scattered among the candidate answers on the n-best list. For example, in response to the open question *What causes RSI?* one candidate answer may be:

> RSI can be caused by repeating the same sequence of movements many times an hour or day.

However, another candidate answer could be:

> RSI is generally caused by a mixture of poor ergonomics, stress and poor posture.

Clearly neither of these two examples constitutes a single complete answer on its own. A more satisfying alternative would require a fusion of the two partial answers into a more complete one such as:

> RSI can be caused by a mixture of poor ergonomics, stress, poor posture and by repeating the same sequence of movements many times an hour or day.

This notion of *sentence fusion* was first introduced by Barzilay (2003). Sentence fusion is a text-to-text generation application, which given two related sentences, outputs a single sentence expressing the information shared by the two input sentences. The process of sentence fusion comprises four major steps:

1. **Linguistic analysis** The input sentences are tokenised and syntactically parsed.
2. **Alignment** The syntax trees are aligned by matching syntactic nodes with similar meaning.
3. **Merging** The syntax trees are combined into a single tree.
4. **Generation** The surface string for the output sentence is generated.

This was originally applied in the context of multi-document summarisation, where it was used to combine similar sentences extracted from different documents in order to increase compression and avoid redundancy. However, as proposed earlier (Marsi and Krahmer, 2005a,b), sentence fusion may be applied in a question-answering context, as well as for the purpose of combining candidate answers.

Ideally, an off-the-shelf sentence fusion system would be used, plugged into an existing QA system, and some evaluations run in order to measure its contribution to the overall answer quality. In reality, there are no readily available sentence fusion systems yet, whereas there are several open research questions. The work reported here reviews the earlier work on sentence fusion. It concerns the implementation and evaluation of models for alignment, merging and generation in Dutch. Section 3.1 starts with the basic question whether it is possible at all to reliably align sentences. After defining the alignment task, the construction of a parallel monolingual corpus consisting of manually aligned syntax trees is described, discussing results on inter-annotator agreement. Section 3.2 then addresses automatic alignment. The algorithm for automatic tree alignment is presented as well as its evaluation on the parallel corpus. Next, Section 3.3 describes exploratory work on simple methods for the merging and generation steps in the sentence fusion process, including an evaluation test. It ends with a discussion and description of more recent follow-up research in Section 3.4.

3.1 Data Collection and Annotation

3.1.1 General Approach to Alignment

Alignment has become standard practice in data-driven approaches to machine translation (e.g. Och and Ney, 2000). Initially work focused on word-based alignment, but more recent research also addresses alignment at the higher levels (substrings, syntactic phrases or trees), e.g., Gildea (2003). The latter approach seems most suitable for current purposes, where the expression is that a sequence of words in one sentence is related to a non-identical sequence of words in another sentence (a paraphrase, for instance). However, if the alignment of arbitrary substrings of two sentences was allowed, then the number of possible alignments grows exponentially to the number of tokens in the sentences, and the process of alignment – either manually or automatically – may become infeasible. An alternative, which seems to occupy the middle ground between word alignment on the one hand and alignment of arbitrary substrings on the other, is to align syntactic analyses. Here, following Barzilay (2003), sentences are aligned at the level of **dependency structures**.

Rather than a binary choice (align or not), one might want to distinguish more fine-grained relations, such as overlap (if two phrases share some, but not all of their content), paraphrases (if two phrases express the same information in different ways), entailments (if one phrase entails the other, but not vice versa), etc. Unlike Barzilay (2003), therefore, not only are similar nodes aligned, but also the node alignments are labelled according to a small set of semantic similarity relations, which will be defined in the next section. This additional information allows for alternative ways of fusing sentences, as described in Section 3.3, which are particularly interesting in the context of QA.

3.1.2 Task Definition

A dependency analysis of a sentence S yields a labelled directed graph $D = \langle V, E \rangle$, where V (vertices) are the nodes, and E (edges) are the dependency relations. For each node v in the dependency structure for a sentence S is defined $\textsc{str}(v)$ as the substring of all tokens under v (i.e., the composition of the tokens of all nodes reachable from v). For example, the string associated with node *persoon* in Figure 2 is *heel veel serieuze personen* (‘very many serious people’).

An alignment between sentences S and S' pairs of nodes from the dependency graphs for both sentences. Aligning node v from the dependency graph D of sentence S with node v' from the graph D' of S' indicates that there is a relation

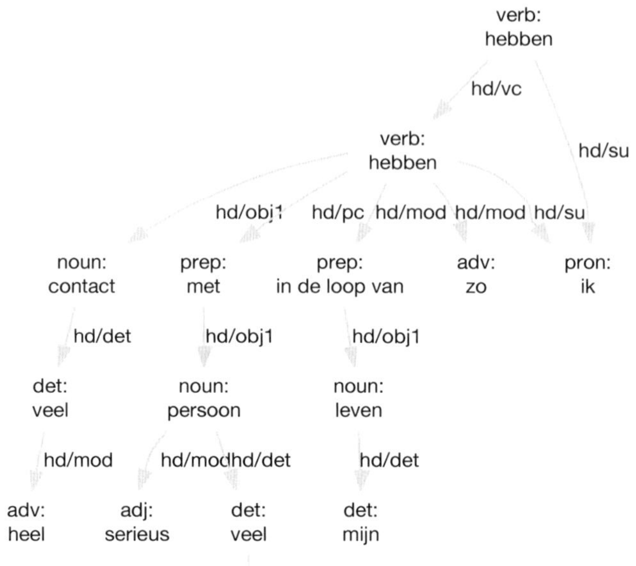

Fig. 2 Example dependency structure for the sentence *Zo heb ik in de loop van mijn leven heel veel contacten gehad met heel veel serieuze personen.* (lit. 'Thus have I in the course of my life had very many contacts with very many serious people').

between $\textsc{str}(v)$ and $\textsc{str}(v')$, i.e., between the respective substrings associated with v and v'. Five potential, mutually exclusive, relations between nodes (with illustrative examples) are distinguished:

1. v **equals** v' iff $\textsc{str}(v)$ and $\textsc{str}(v')$ are literally identical (abstracting from case and word order)
   Example: "a small and a large boa constrictor" equals "a large and a small boa constrictor";
2. v **restates** v' iff $\textsc{str}(v)$ is a paraphrase of $\textsc{str}(v')$ (same information content but different wording),
   Example: "a drawing of a boa constrictor snake" restates "a drawing of a boa constrictor";
3. v **specifies** v' iff $\textsc{str}(v)$ is more specific than $\textsc{str}(v')$,
   Example: "the planet B 612" specifies "the planet";
4. v **generalises** v' iff $\textsc{str}(v')$ is more specific than $\textsc{str}(v)$,
   Example: "the planet" generalises "the planet B 612";
5. v **intersects** v' iff $\textsc{str}(v)$ and $\textsc{str}(v')$ share some informational content, but also each express some piece of information not expressed in the other,
   Example: "Jupiter and Mars" intersects "Mars and Venus"

Note that there is an intuitive relation with entailment here: both *equals* and *restates* can be understood as mutual entailment (i.e., if the root nodes of the analyses corresponding S and S' stand in an equal or restate relation, S entails S' and S' entails S), if S *specifies* S' then S also entails S' and if S *generalises* S' then S is entailed by S'.

An alignment between S and S' can now formally be defined on the basis of the respective dependency graphs $D = \langle V, E \rangle$ and $D' = \langle V', E' \rangle$ as a graph $A = \langle V_A, E_A \rangle$, such that

$$E_A = \{\langle v, l, v' \rangle \mid v \in V \ \& \ v' \in V' \ \& \ l(\text{STR}(v), \text{STR}(v'))\},$$

where l is one of the five relations defined above. The nodes of A are those nodes from D en D' which are aligned, formally defined as

$$V_A = \{v \mid \exists v' \exists l \langle v, l, v' \rangle \in E_A\} \cup \{v' \mid \exists v \exists l \langle v, l, v' \rangle \in E_A\}$$

3.1.3 Corpus

For evaluation and parameter estimation a **parallel monolingual corpus** was developed, consisting of two different Dutch translations of the French book "Le petit prince" (*the little prince*) by Antoine de Saint-Exupéry (published 1943), one by Laetitia de Beaufort-van Hamel (1966) and one by Ernst van Altena (2000). The texts were automatically tokenised and split into sentences, after which errors were manually corrected. Corresponding sentences from both translations were manually aligned; in most cases this was a one-to-one mapping, but occasionally a single sentence in one version mapped onto two sentences in the other. Next, the **Alpino** parser for Dutch (e.g. Bouma et al, 2001) was used for part-of-speech tagging and lemmatising all words, and for assigning a dependency analysis to all sentences. The POS labels indicate the major word class (e.g. *verb*, *noun*, *pron*, and *adv*). The dependency relations hold between tokens and are the same as used in the Spoken Dutch Corpus (see e.g. van der Wouden et al, 2002). These include dependencies such as *head/subject*, *head/modifier* and *coordination/conjunction*. See Figure 2 as an example. If a full parse could not be obtained, Alpino produced partial analyses collected under a single root node. Errors in lemmatisation, POS-tagging, and syntactic dependency parsing were not subject to manual correction.

3.1.4 Results

All text material was aligned by two annotators. They started by aligning the first ten sentences of Chapter One together in order to get a feel for the task. They continued with the remaining sentences from Chapter One individually. The total number of nodes in the two translations of the chapter was 445 and 399 respectively. Inter-annotator agreement was calculated for two aspects: alignment and relation labelling. With respect to alignment, the precision was calculated, recalled and F-scored (with $\beta = 1$) on aligned node pairs as follows:

$$precision(A_{real}, A_{pred}) = \frac{|A_{real} \cap A_{pred}|}{|A_{pred}|}$$

$$recall(A_{real}, A_{pred}) = \frac{|A_{real} \cap A_{pred}|}{|A_{real}|}$$

$$F\text{-}score = \frac{2 \times precision \times recall}{precision + recall}$$

where A_{real} is the set of all real alignments (the reference or golden standard), A_{pred} is the set of all predicted alignments, and $A_{pred} \cap A_{real}$ is the set of all correctly predicted alignments. For the purpose of calculating inter-annotator agreement, one of the annotations (A_1) was considered the 'real' alignment, the other (A_2) the 'predicted'. The results are summarised in Table 5 in column (A_1, A_2).

Table 5 Inter-annotator agreement with respect to alignment between annotators 1 and 2 before (A_1, A_2) and after $(A_{1'}, A_{2'})$ revision, and between the consensus and annotator 1 $(A_c, A_{1'})$ and annotator 2 $(A_c, A_{2'})$ respectively.

	(A_1,A_2)	$(A_{1'},A_{2'})$	$(A_c,A_{1'})$	$(A_c,A_{2'})$
#real:	322	323	322	322
#pred:	312	321	323	321
#correct:	293	315	317	318
Precision:	.94	.98	.98	.99
Recall:	.91	.98	.98	.99
F-score:	.92	.98	.98	.99

Next, both annotators discussed the differences in alignment, and corrected mistaken or forgotten alignments. This improved their agreement, as shown in column $(A_{1'}, A_{2'})$. In addition, they agreed on a single consensus annotation (A_c). The last two columns of Table 5 show the results of evaluating each of the revised annotations against this consensus annotation. The F-score of .96 can therefore be regarded as the upper bound on the alignment task.

Table 6 Inter-annotator agreement with respect to alignment **relation labelling** between annotators 1 and 2 before (A_1, A_2) and after $(A_{1'}, A_{2'})$ revision, and between the consensus and annotator 1 $(A_c, A_{1'})$ and annotator 2 $(A_c, A_{2'})$ respectively.

	(A_1,A_2)	$(A_{1'},A_{2'})$	$(A_c,A_{1'})$	$(A_c,A_{2'})$
Precision:	.86	.96	.98	.97
Recall:	.86	.95	.97	.97
F-score:	.85	.95	.97	.97
κ:	.77	.92	.96	.96

In a similar way, the agreement was calculated for the task of labelling the alignment relations. Results are shown in Table 6, where the measures are *weighted* precision, recall and F-score. For instance, the precision is the weighted sum of the separate precision scores for each of the five relations. The table also shows the κ-score, which is another commonly used measure for inter-annotator agreement (Carletta, 1996). Again, the F-score of .97 can be regarded as the upper bound on the relation labelling task.

It is felt that these numbers indicate that the labelled alignment task is well defined and can be accomplished with a high level of inter-annotator agreement.

3.2 Automatic Alignment

3.2.1 Tree Alignment Algorithm

The tree alignment algorithm is based on Meyers et al (1996), and similar to that used in Barzilay (2003). It calculates the match between each node in **dependency tree** D against each node in dependency tree D'. The score for each pair of nodes only depends on the similarity of the words associated with the nodes and, recursively, on the scores of the best matching pairs of their descendants. For an efficient implementation, dynamic programming is used to build up a score matrix, which guarantees that each score will be calculated only once.

Given two dependency trees D and D', the algorithm builds up a score function $S(v,v')$ for matching each node v in D against each node v' in D', which is stored in a matrix M. The value $S(v,v')$ is the score for the best match between the two subtrees rooted at v in D and at v' in D'. When a value for $S(v,v')$ is required, and is not yet in the matrix, it is recursively computed by the following formula:

$$S(v,v') = max \begin{cases} \text{TREEMATCH}(v,v') \\ max_{i=1,\ldots,n}\ S(v_i,v') \\ max_{j=1,\ldots,m}\ S(v,v'_j) \end{cases}$$

where $v_1,\ldots,v_n$ denote the children of v and $v'_1,\ldots,v'_m$ denote the children of v'. The three terms correspond to the three ways that nodes can be aligned: (1) v can be directly aligned to v'; (2) any of the children of v can be aligned to v'; (3) v can be aligned to any of the children of v'. Notice that the last two options imply skipping one or more edges, and leaving one or more nodes unaligned.[3]

The function $\text{TREEMATCH}(v,v')$ is a measure of how well the subtrees rooted at v and v' match:

$$\text{TREEMATCH}(v,v') = \text{NODEMATCH}(v,v') + \\ \max_{p\in\mathscr{P}(v,v')} \left[\sum_{(i,j)\in p} \left(\text{RELMATCH}(\overrightarrow{v}_i, \overrightarrow{v}'_j) + S(v_i,v'_j) \right) \right]$$

[3] In the original formulation of the algorithm by Meyers et al (1996), there is a penalty for skipping edges.

Here $\vec{v}_i$ denotes the dependency relation from v to v_i. $\mathscr{P}(v,v')$ is the set of all possible pairings of the n children of v against the m children of v', which is the power set of $\{1,\ldots,n\} \times \{1,\ldots,m\}$. The summation ranges over all pairs, denoted by (i,j), which appear in a given pairing $p \in \mathscr{P}(v,v')$. Maximising this summation thus amounts to finding the optimal alignment of children of v to children of v'.

$\textsc{NodeMatch}(v,v') \geq 0$ is a measure of how well the label of node v matches the label of v'.

$\textsc{RelMatch}(\vec{v}_i, \vec{v}'_j) \geq 0$ is a measure for how well the dependency relation between node v and its child v_i matches that of the dependency relation between node v' and its child v_j.

Since the dependency graphs delivered by the Alpino parser were usually not trees, they required some modification in order to provide suitable input for the tree alignment algorithm. First, a root node was determined, which is defined as a node from which all other nodes in the graph can be reached. In the rare case of multiple root nodes, an arbitrary one was chosen. Starting from this root node, any cyclic edges were temporarily removed during a depth-first traversal of the graph. The resulting directed acyclic graphs may still have some degree of structure sharing, but this poses no problem for the algorithm.

3.2.2 Evaluation of Automatic Alignment

The automatic alignment of nodes was evaluatd, abstracting from relation labels, as there is no algorithm for automatic labelling of these relations yet. The baseline is achieved by aligning those nodes which stand in an *equals* relation to each other, i.e., a node v in D is aligned to a node v' in D' iff $\textsc{str}(v) = \textsc{str}(v')$. This alignment can be constructed relatively easily.

The alignment algorithm is tested with the following NODEMATCH function:

$$\textsc{NodeMatch}(v,v') = \begin{cases} 10 & \text{if } \textsc{str}(v) = \textsc{str}(v') \\ 5 & \text{if } \textsc{label}(v) = \textsc{label}(v') \\ 2 & \text{if } \textsc{label}(v) \text{ is a synonym} \\ & \text{ hyperonym or hyponym} \\ & \text{of } \textsc{label}(v') \\ 0 & \text{otherwise} \end{cases}$$

It reserves the highest value for a literal string match, a somewhat lower value for matching lemmas, and an even lower value in case of a synonym, hyperonym or hyponym relation. The latter relations are retrieved from the Dutch part of EuroWordnet (Vossen, 1998). For the RELMATCH function, simply a value of 1 was used for identical dependency relations, and 0 otherwise. These values were found to be adequate in a number of test runs on two other, manually aligned chapters (these chapters were not used for the actual evaluation). The intention is to to experiment with automatic optimisations in the future.

The alignment accuracy was measured and defined as the percentage of correctly aligned node pairs, where the consensus alignment of Chapter One served as the golden standard. The results are summarised in Table 7. In order to test the contribution of synonym and hyperonym information for node matching, performance is measured with and without the use of EuroWordnet. The results show that the algorithm improves substantially on the baseline. The baseline already achieves a relatively high score (an F-score of .56), which may be attributed to the nature of the material: the translated sentence pairs are relatively close to each other and may show a sizeable amount of literal string overlap. The alignment algorithm (without use of EuroWordnet) loses a few points on precision, but improves considerably on recall (a 200% increase with respect to the baseline), which in turn leads to a substantial improvement on the overall F-score. The use of EuroWordnet leads to a small increase (two points) on both precision and recall (and thus to a small increase on F-score). Yet, in comparison with the gold standard human score for this task (.95), there is clearly room for further improvement.

Table 7 Precision, recall and F-score on automatic alignment.

Alignment :	*Prec* :	*Rec* :	*F-score:*
Baseline	.87	.41	.56
Algorithm without wordnet	.84	.82	.83
Algorithm with wordnet	.86	.84	.85

3.3 Merging and Generation

The remaining two steps in the sentence fusion process are merging and generation. In general, **merging** amounts to deciding which information from either sentence should be preserved, whereas **generation** involves producing a grammatically correct surface representation. In order to get an idea about the baseline performance, a simple, somewhat naive string-based approach, was explored. Below, the pseudo code is shown for merging two dependency trees in order to get restatements. Given a labelled alignment A between dependency graphs D and D', if there is a **restates** relation between node v from D and node v' from D', the string realisation of v' was added as an alternative to those of v.

RESTATE(A)
for each edge $\langle v, l, v' \rangle \in E_A$
  do if $l =$ **restates**
    then $\text{STR}(v) \leftarrow \text{STR}(v) \vee \text{STR}(v')$

The same procedure is followed in order to get specifications:

SPECIFY(A)

for each edge $\langle v, l, v' \rangle \in E_A$
  do if $l =$ **generalizes**
    then $\text{STR}(v) \leftarrow \text{STR}(v) \vee \text{STR}(v')$

The generalisation procedure adds the option to omit the realisation of a modifier that is *not* aligned:

GENERALIZE(D,A)

for each edge $\langle v, l, v' \rangle \in E_A$
  do if $l =$ **specifies**
    then $\text{STR}(v) \leftarrow \text{STR}(v) \vee \text{STR}(v')$
for each edge $\langle v, l, v' \rangle \in E_D$
  do if $l \in$ MOD-DEP-RELS and $v \notin E_A$
    then $\text{STR}(v) \leftarrow \text{STR}(v) \vee \text{NIL}$

where MOD-DEP-REL is the set of dependency relations between a node and a modifier (e.g. *head/mod* and *head/predm*).

Each procedure is repeated twice, once adding substrings from D into D', and once the other way around. Next, the dependency trees are traversed and all string realisations are generated, extending the list of variants for each node that has multiple realisations. Finally, multiple copies of the same string are filtered out, as well as strings that are identical to the input sentences.

As expected, many of the resulting variants are ungrammatical, because constraints on word order, agreement or subcategorisation are violated. Following work on statistical surface generation (Langkilde and Knight, 1998) and other work on sentence fusion (Barzilay, 2003), we try to filter out ungrammatical variants with an n-gram language model. The Cambridge-CMU Statistical Modelling Toolkit v2 was used to train a 3-gram model on over 250M words from the Twente Newscorpus, using back-off and Good-Turing smoothing. Variants were ranked in order of increasing entropy.

To gain some insight into the general performance of the merging and generation strategy, a small evaluation test was performed in which two judges independently judged all generated variants in terms of three categories:

1. **Perfect**: no problems in either semantics or syntax;
2. **Acceptable**: understandable, but with some minor flaws in semantics or grammar;
3. **Nonsense**: serious problems in semantics or grammar.

Table 8 shows the number of sentences in each of the three categories per judge, broken down in restatements, generalisation and specifications. The κ-score on this classification task is .75, indicating a moderate to good agreement between the judges. Roughly half of the generated sentences are perfect, although specifications are somewhat less well-formed.

The conclusion from this evaluation is that sentence fusion is a viable and interesting approach for producing restatements, generalisation and specifications.

Table 8 Results of the evaluation of the sentence fusion output as the number of sentences in each of the three categories *perfect*, *acceptable* and *nonsense* per judge (J1 and J2), broken down in restatements, generalisations and specifications.

	Restate		Specific		General	
	J1	J2	J1	J2	J1	J2
Perfect:	109	104	28	22	89	86
Acceptable:	44	58	15	16	34	24
Nonsense:	41	32	19	24	54	67
Total:	194		62		177	

However, there is certainly further work to do; the procedure for merging dependency graphs should be extended, and the realisation model clearly requires more linguistic sophistication, in particular to deal with word order, agreement and subcategorisation constraints.

3.4 Discussion

The early work on sentence fusion described here was only performed on a small corpus, consisting of parallel translations of a single book, (*Le Petit Prince*). An evaluation of sentence fusion in the context of QA has not been offered. Both limitations have been addressed in the DAESO project (short for Detecting and Exploiting Semantic Overlap), which was a partial continuation of the IMOGEN project. In this project, a one million word parallel monolingual treebank was developed (Marsi and Krahmer, 2007, 2009), containing multiple translations of various books (*Le Petit Prince*, but also parts of Darwin's *Origin of Species* and Montaigne's *Essays*), as well as multiple news reports about the same event, headlines for related news reports and multiple answers to the same question (this data was actually collected in the IMIX project).

Half of the data was manually annotated, for which two annotation tools were developed (both publicly available): Hitaext, allowing many-to-many alignments on the sentence level, and Algraeph, for sentence alignment at the word and phrase level.[4] The other half of the corpus was automatically aligned, with a vastly improved version of the automatic aligner described above. One important aspect of the automatic aligner is that it makes no prior assumptions about the relation between the two sentences other than that they are somehow related: the amount of overlap may range from a few words (as may be the case in the news reports) to the entire sentence (as is the case in the parallel translations), and no order between the sentences is assumed. Evaluation (on alignment of news reports) shows that the performance of the automatic aligner approaches that of the human annotators, where it is interesting to observe that the algorithm outperforms human annotators on the Equals and Intersects relations. Human annotators, by contrast, are better at

[4] Both tools are available from `http://daeso.uvt.nl`

classifying the remaining relations, but since Equals and Intersects are relatively frequent in the News segment, the overall weighted performance of the algorithm is less than a percent below the scores obtained by the human annotators (Marsi and Krahmer, 2010).

DAESO also continues the work on sentence fusion. It has been argued that sentence fusion is a poorly defined task, which is therefore difficult to evaluate (Daumé III and Marcu, 2004). Krahmer et al (2008) studied sentence fusion in a QA setting, where participants were asked to merge potential answer sentences with and without explicitly showing the question. Question-based sentence fusion was found to be a better defined task than generic sentence fusion (Q-based fusions are shorter, display less variety in length, yield more identical results and have higher normalised Rouge scores). The sentence fusion data collected in this way has been made publicly available. Moreover, it was discovered that in a QA application, participants strongly prefer Q-based fusions to generic ones, and have a preference for union fusions (combining information from answers, without overlap) over intersection fusions (only using the shared information in potential answer sentences). This clearly shows that sentence fusion is indeed a useful strategy for QA systems.

4 Conclusion

This chapter describes two related ways in which QA systems could provide more informative answers, by (1) doing query-based summarisation and (2) fusing potential answer sentences to more complete answers.

Concerning query-based summarisation, the primary aim was to bring automatic content selection practice in line with insights from discourse theory. To this end, a framework for automatic summarisation was devised which was founded on graph theory and can be applied as a text-to-text generation technique in Question Answering. The content selection algorithm is entirely based on the relations between content units (text passages). The evaluated systems are just examples of possible implementations of this framework; they can be extended to exploit more textual features, and discourse oriented features in particular.

The framework represents a step towards context aware summarisation. Previous work on query-based summarisation has mainly focused on extracting the set of sentences which best match the query, ignoring their broader context. The features used for relating sentences are computationally low-cost and easy to port to other languages, but knowledge-intensive methods may detect relations between sentences more accurately. Despite this, the graph-based approach showed good results compared to the DUC participant systems (the redundancy-aware probabilistic relevance system would have ranked first for Rouge-2 and second for Rouge-SU4 if it had participated in DUC 2005, when most of the work described in this chapter was done), which indicates that we are on the right track. The main lessons learned from the experiments are the following:

1. The graph-based approach to summarisation represents a promising direction for further research, given the good results in spite of the superficial linguistic analysis performed by the evaluated systems. Even better results are to be expected when more sophisticated features are used.
2. The probabilistic interpretation of semantic networks (i.e., *probabilistic relevance*) seems to be more suitable for content selection than the social network interpretation (i.e., *normalised centrality*).

Further performance gains may be achieved by using more different sources of information for detecting relations, including knowledge-intensive methods such as rhetorical relation detection or anaphora resolution.

This chapter also described the first explorations into sentence fusion for Dutch. The starting point was the sentence fusion model proposed by Barzilay et al (1999); Barzilay (2003), and further extended by Barzilay and McKeown (2005), in which dependency analyses of pairs of sentences are first aligned, after which the aligned parts (representing the common information) are fused. The resulting fused dependency tree is subsequently transferred into natural language. The new contributions are primarily in two areas: First, an explicit evaluation of the alignment was carried out – both human and automatic alignment – whereas Barzilay (2003) only evaluates the output of the complete sentence fusion process. The annotators can reliably align phrases and assign relation labels to them, and that good results can be achieved with automatic alignment, certainly above an informed baseline, albeit still below human performance. Second, Barzilay and her colleagues developed the sentence fusion model in the context of multi-document summarisation, but arguably the approach could also be used for applications such as QA or information extraction. This seems to call for a more refined version of sentence fusion, which has consequences for alignment, merging and realisation. Therefore, five different types of semantic relations between strings (i.e. equals, restates, specifies, generalises and intersects) have been introduced. This increases the expressiveness of the representation, and supports generating restatements, generalisations and specifications. The first results on sentence realisation were described and evaluated based on these refined alignments, with promising results.

References

Bakshi K, Huynh D, Katz B, Karger D, Lin J, Quan D, Sinha V (2003) The role of context in question answering systems. In: CHI '03 extended abstracts on Human Factors in Computing Systems, New York, NY, USA, pp 1006–1007

Barzilay R (2003) Information fusion for multidocument summarization. PhD thesis, Columbia University

Barzilay R, Elhadad M (1997) Using lexical chains for text summarization. In: Proceedings of the ACL workshop on Intelligent Scalable Text Summarization, pp 10–17

Barzilay R, McKeown K (2005) Sentence fusion for multidocument news summarization. Computational Linguistics 31(3):297–328

Barzilay R, McKeown K, Elhaded M (1999) Information fusion in the context of multi-document summarization. In: Proceedings of the 37th annual meeting of the ACL, Maryland

Bates M (1990) The berry-picking search: user interface design. In: Thimbleby H (ed) User Interface Design, Addison-Wesley

Blair-Goldensohn S, McKeown K (2006) Integrating rhetorical-semantic relation models for query-focused summarization. In: Proceedings of the Document Understanding Conference

Bouma G, van Noord G, Malouf R (2001) Alpino: Wide-coverage computational analysis of Dutch. In: Proceedings of CLIN

Carbonell J, Goldstein J (1998) The use of MMR, diversity-based reranking for reordering documents and producing summaries. In: Proceedings of the 21st Annual International ACM SIGIR Conference on Research and Development in Information Retrieval, New York, NY, USA, pp 335–336

Carletta J (1996) Assessing agreement on classification tasks: the kappa statistic. Compututational Linguistics 22(2):249–254

Daumé III H, Marcu D (2004) Generic sentence fusion is an ill-defined summarization task. In: Proceedings of the ACL workshop: Text Summarization Branches Out, Barcelona, Spain

Edmundson HP (1969) New methods in automatic extracting. Journal of the ACM 16(2):264–285

Erkan G, Radev D (2004) LexRank: Graph-based centrality as salience in text summarization. Journal of Artificial Intelligence Research

Gildea D (2003) Loosely tree-based alignment for machine translation. In: Proceedings of the 41st annual meeting of the ACL, Sapporo, Japan

Krahmer E, Marsi E, van Pelt P (2008) Query-based sentence fusion is better defined and leads to more preferred results than generic sentence fusion. In: Proceedings of the 46th Annual Meeting of the ACL, Columbus, OH, USA, pp 193–196

Langkilde I, Knight K (1998) Generation that exploits corpus-based statistical knowledge. In: Proceedings of the 36th annual meeting of the ACL, Morristown, NJ, USA, pp 704–710

Lin CY (2004) Rouge: a package for automatic evaluation of summaries. In: Proceedings of the ACL workshop: Text Summarization Branches Out, Barcelona, Spain

Luhn H (1958) The automatic creation of literature abstracts. IBM Journal of Research and Development 2(2):159–165

Mani I, Bloedorn E (1997) Multi-document summarization by graph search and matching. In: Proceedings of the Fourteenth National Conference on Artificial Intelligence, pp 622–628

Mann W, Thompson S (1988) Rhetorical Structure Theory: Toward a functional theory of text organization. Text 8:243–281

Marcu D (1999) Discourse trees are good indicators of importance in text. In: Mani I, Maybury M (eds) Advances in Automatic Text Summarization, MIT Press, pp 123–136

Marsi E, Krahmer E (2005a) Classification of semantic relations by humans and machines. In: Proceedings of the ACL workshop on Empirical Modeling of Semantic Equivalence and Entailment, Ann Arbor, Michigan, pp 1–6

Marsi E, Krahmer E (2005b) Explorations in sentence fusion. In: Proceedings of the 10th European workshop on Natural Language Generation, Aberdeen, UK

Marsi E, Krahmer E (2007) Annotating a parallel monolingual treebank with semantic similarity relations. In: Proceedings of the 6th International Workshop on Treebanks and Linguistic Theories, Bergen, Norway, pp 85–96

Marsi E, Krahmer E (2009) Detecting semantic overlap: A parallel monolingual treebank for dutch. In: Proceedings of CLIN

Marsi E, Krahmer E (2010) Automatic analysis of semantic similarity in comparable text through syntactic tree matching. In: Proceedings of the 23rd International Conference on Computational Linguistics, Beijing, China, pp 752–760

Maybury M (2004) New Directions in Question Answering. AAAI Press

Meyers A, Yangarber R, Grisham R (1996) Alignment of shared forests for bilingual corpora. In: Proceedings of 16th International Conference on Computational Linguistics, Copenhagen, Denmark, pp 460–465

Noreen EW (1989) Computer intensive methods for testing hypotheses: an introduction. Wiley, New York, NY, USA

Och FJ, Ney H (2000) Statistical machine translation. In: EAMT Workshop, Ljubljana, Slovenia, pp 39–46

Porter M (2001) Snowball: A language for stemming algorithms. Http://snowball.tartarus.org/texts/introduction.html

Spärck Jones K (1972) A statistical interpretation of term specificity and its application in retrieval. Journal of Documentation 28(1):11–21

Strzalkowski T, Gaizauskas R, Voorhees E, Harabagiu S, Weishedel R, Israel D, Jacquemin C, Lin C, Maiorano S, Miller G, Moldovan D, Ogden B, Prager J, Riloff E, Burger J, Singhal A, Cardie C, Shrihari R, Chaudhri V (2000) Issues, tasks, and program structures to roadmap research in question & answering (Q&A). NIST

Vossen P (ed) (1998) EuroWordNet: a multilingual database with lexical semantic networks. Kluwer Academic Publishers, Norwell, MA, USA

Wolf F, Gibson E (2005) Representing discourse coherence: A corpus-based study. Computational Linguistics 31(2):249–288

van der Wouden T, Hoekstra H, Moortgat M, Renmans B, Schuurman I (2002) Syntactic analysis in the spoken dutch corpus. In: Proceedings of the 3rd International Conference on Language Resources and Evaluation, Las Palmas, Spain, pp 768–773

Part IV
Text Analysis for Question Answering

Automatic Extraction of Medical Term Variants from Multilingual Parallel Translations

Lonneke van der Plas, Jörg Tiedemann and Ismail Fahmi

Abstract The extraction of terms and their variants is an important issue in various applications of natural language processing (NLP), such as question answering and information retrieval. This chapter discusses a method to automatically extract medical terms and their variants from a multilingual corpus of parallel translations. As a first step terms are extracted using a pattern-based approach. In order to determine what terms are variants of each other the distributional method used calculates semantic similarity between terms on the basis of translations of these terms in multiple languages. Word alignment techniques were used in combination with phrase extraction techniques from phrase-based machine translation to extract phrases and their translations from a medical parallel corpus. The approach provides a promising strategy for the extraction of term variants using straightforward and fully-automatic techniques. Moreover, the approach is independent of domain and language and can thus be applied to various domains and various languages for which parallel multilingual corpora exist.

1 Introduction

Within the framework of the IMIX (Interactive Multimodal Information eXtraction) project, the QADR team (Bouma et al, 2007) developed a question answering (QA) system for Dutch that relies heavily on linguistic information. In the course of the

Lonneke van der Plas
Department of Linguistics, University of Geneva, Geneva, Switzerland,
e-mail: lonneke.vanderplas@unige.ch

Jörg Tiedemann
Department of Linguistics and Philology, Uppsala University, Uppsala, Sweden,
e-mail: jorg.tiedemann@lingfil.uu.se

Ismail Fahmi
Gresnews Media, Amsterdam, The Netherlands, e-mail: ismail.fahmi@gmail.com

research it was noticed that to be able to return adequate questions to users one needed to be able to deal with variation in terminology between the term in the question and the terms used in a document that contains the answer. An example taken from the CLEF evaluation forum is given below:

Waar explodeerde de eerste atoombom?
'Where did the first nuclear bomb explode?'

The answer was found in a document that contained the word *ontploffen* 'to blow up'. However, it did not contain the original term used in the question *exploderen* 'to explode'.

People use a wide variety of terms to describe the same concept. Furnas et al (1987) have shown in several naming experiments with domain specialists and novices that the probability that people will use the same term to describe the same object is less than 20%. So if a question answering system relies soley on matching words from the user's query with words in the document, it runs a very high risk of missing relevant documents, due to differences in wording.

The IMIX project is mainly concerned with question answering for the medical domain. The problems related to term variation manifest themselves particularly clearly in the medical domain, where the terms used by medical specialists and laymen are very different, but the two parties are in frequent communication. This important terminological gap explains the interest in building terminological resources for the medical domain. The Unified Medical Language System (McCray and Hole, 1990), for example, comprises a large database that consists of three knowledge sources in the domain of health and biomedicine, the Metathesaurus, the Semantic Network and the SPECIALIST Lexicon. The IMIX project builds question answering systems for Dutch and would therefore benefit from lexical resources for the Dutch medical domain. Fahmi (2009) shows that the lack of term normalisation is one of the largest sources of errors for the Dutch medical QA system developed by the QADR team. From the total of 17 incorrectly answered questions, 29% of them are caused by normalisation problems either in the question analysis part or the answer extraction part. These problems include keyword capitalisation in the questions, diacritics, non-term head words, and synonym. Although the UMLS has some multilingual parts, the number of resources available for languages other than English is very limited. This is why a method was proposed to acquire medical term variants for Dutch automatically.

If multiple words refer to the same concept, there exists a synonym relation between these words: The words have the same meaning. The synonym relation is not the only relation that holds between words. For example, there exists a relation of mere association between the words *night* and *darkness* and there exists a hypernym or ISA relation between *animal* and the subordinate term *dog*. The automatic acquisition of hypernyms is investigated in the chapter by Tjong Kim Sang, Hofmann, and de Rijke, *Extraction of Hypernymy Information from Text* (this volume).

Because these relations hold between words (the lexical elements) by virtue of their meaning (semantics), this type of information is referred to as lexico-semantic knowledge. Kilgarriff and Yallop (2000) use the terms *loose* and *tight* to describe different types of lexico-semantic resources. In a *loose* resource, such as the Roget Thesaurus (Roget, 1911), words are related in an associative way. They are related according to the subject field, whereas *tight* resources tend to group words that are the same kind of things, i.e. that belong to the same semantic class, together. Princeton WordNet (Fellbaum, 1998) is an electronic resource inspired by current psycholinguistic theories of human lexical memory. Synonyms are grouped in synsets, i.e. lists of words that are (near)-synonyms. These synsets are in turn related by basic semantic relations. Synonymy expresses a tight relationship between words. In the above example synonymy information was needed to answer the question adequately.

This chapter describes a method to acquire lexico-semantic information, in particular synonyms, automatically using a distributional method. It builds on the idea that semantically related words are distributed similarly over contexts (distributional hypothesis). Harris (1968) claims that, 'the meaning of entities and the meaning of grammatical relations among them, is related to the restriction of combinations of these entities relative to other entities.' This is in line with the Firthian saying that, 'You shall know a word by the company it keeps.' (Firth, 1957). In other words, you can grasp the meaning of a word by looking at its contexts.

The context can be defined in many ways. For example, the verbs that are in a subject relation with a particular noun form a part of its context. Previous work has been mainly concerned with the syntactic contexts a word is found in (Lin, 1998; Curran, 2003). These contexts can be used to determine the semantic relatedness of words. For instance, words that occur in a object relationship with the verb *to peel* have something in common: they have a skin, rind, bark or other covering. Other work has been concerned with the bag-of-word context (Wilks et al, 1993; Schütze, 1992), where the context of a word is the words that are found in its proximity. Yet another context is the translational context.

Van der Plas (2008a) has shown that the way the context is defined influences the type of lexico-semantic knowledge that is discovered. After gold standard evaluations and manual inspection van der Plas (2008a) concludes that the alignment-based method is better at finding synonyms than the syntax-based method. The performance of the former is almost twice as good as the latter, while the amount of data used is much smaller. The poorer performance was ascribed to the syntax-based method to the fact that loosely related words, such as *wine* and *beer*, are often found in the same syntactic contexts. The alignment-based method suffers less from this indiscriminant acceptance. Words are typically translated by words with the same meaning. The word *wine* is typically not translated with a word for *beverage* nor with a word for *beer*, and neither is *good* translated with a word for *bad*.

This chapter will be concerned with the translational context. The translational context of a word is the set of translations it gets in other languages. For example, the translational context of *cat* is *kat* in Dutch and *chat* in French. This requires a rather broad understanding of the term context.

When translations for words are needed, bilingual dictionaries seem to be a straightforward place to start looking. However, dictionaries are static and, therefore, often incomplete resources, and the majority of dictionaries do not provide frequency information. Also, they are not always publicly available for all languages. Consequently, it was decided to automatically acquire word translations in multiple languages from text. Text in this case should be understood to mean multilingual parallel text. Automatic alignment gives the translations of a word in multiple languages. Any multilingual parallel corpus can be used for this purpose. It is thus possible to focus on a special domain by selecting parallel corpora from a specific domain. For this chapter, a corpus of parallel translations in the medical domain has been selected to extract medical term variants. Furthermore, the automatic alignment provides us with frequency information for every translation pair specific to the domain underconsideration. Thus, the translations a word gets in many languages accompanied by the frequency of the translations gives us a unique mould for each term in the domain under consideration.

How does one get from translational contexts to synonymy? The idea is that words that share a large number of translations are similar. For example both *autumn* and *fall* get the translation *herfst* in Dutch, *Herbst* in German, and *automne* in French. This indicates that *autumn* and *fall* are synonyms.

Aligned parallel corpora have often been used for the task of discriminating the different senses words have (word sense discovery). The idea behind this approach is that a word that receives different translations might be polysemous. For example, a word such as *wood* receives the translation *woud* and *hout* in Dutch. The former refers to an area with many trees and the latter, to the solid material derived from trees. This type of work is all built on the divergence of translational context, i.e. one word in the source language is translated by many different words in the target language. However, here, the convergence of translations is of interest, i.e. two words in the source language receiving the same translation in the target language.

Of course these two phenomena are not independent. The alleged conversion of the target language might well be a hidden diversion of the source language. Since the English word might be polysemous, the fact that *woud*, an area with many trees, and *hout*, the material made from trees, in Dutch are both translated in English by *wood* does not mean that *woud* and *hout* in Dutch are synonyms. However, as shown by van der Plas (2008a) the effects of polysemy can be remedied by using translations from multiple languages at the same time. Languages other than English do not use the same words to translate *woud* and *hout*. In French one find *forêt* for *woud* and *bois* for *hout*.

So far it is explained how to find words that have a similar meaning from corpora of parallel translations. However, this chapter is interested in the extraction of terms for a specific domain: the medical domain. Medical terms are often composed of multiple words, such as *aangeboren afwijking* 'birth defect' for *congenitale aandoening* 'congenital disorder'. Variation in the surface form of these terms is referred to with the word TERM VARIATION. The alignment-based distributional method described in van der Plas (2008a) has been applied to the discovery of single word synonyms for the general domain. In this chapter[1] it is described how the method is used to find multiword terminological variants for the medical domain.

This chapter begins by describing its scope, before moving on to the methodology, and more specifically the types of term variants this methodology is trying to uncover. Ville-Ometz et al (2008) describe a general typology of term variants, including 7 different types: graphic variation (R factor /resistance factor), inflectional variation (number or gender variations), paraphrastic syntactic variations (mechanism of resistance/ resistance mechanism), incremental syntactic variations (epithelial cell / epithelial cultured cell), a combination of the two previous types of variation, morphosyntactic variation, and paradigmatic variation, where one of the content words is substituted by a synonym. The research in this chapter is limited to meaning-preserving term variants. Incremental syntactic variations are outside the scope of this chapter. Furthermore, inflectional variation is not sought. Data worked on has been normalised for inflectional variation. The canonical form of terms has been work on. The application being worked on (the QA system) deals with inflectional variation internally. On the other hand, any kind of variation in terms of synonymy is allowed for. It is not limited to one of the terms as it is in the above description of paradigmatic variation. To summarise, the aim is to extract all kinds of meaning-preserving variation in terminology on the basis of the canonical forms of terms.

Word alignment techniques will be used, in combination with phrase extraction techniques from phrase-based machine translation, to extract phrases and their translations from a medical parallel corpus. Filters based on Part-of-Speech patterns will be applied to extract candidate terms from these phrase tables. Finally, similarity measures from distributional methods are used to compute the list of term variants from this data.

2 Alignment-based Methods

This section explains the alignment-based approaches to distributional similarity. It gives some examples of translational context (2.1) and explains how measures serve to determine the similarity of these contexts (2.2). This section ends with a discussion of related work (2.3).

2.1 Translational Context

The translational context of a word or a multiword term is the set of translations it gets in other languages. There are several ways to get hold of the translational context of words. One is to look up translations in a dictionary. The approach being

[1] Part of the material in this chapter has been published as van der Plas and Tiedemann (2010).

proposed relies on automatic word alignment of parallel corpora from Dutch to one or more target languages.

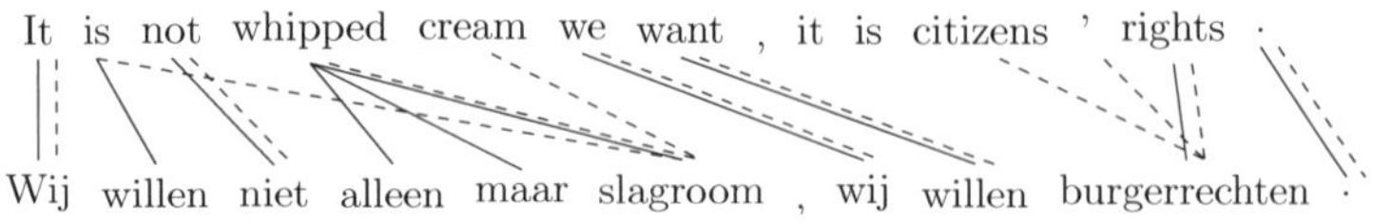

Fig. 1 Example of bidirectional word alignments of two parallel sentences.

Figure 1 illustrates the automatic word alignment between a Dutch and an English phrase, as a result of using the IBM alignment models (Brown et al, 1993). The alignment of two texts is bi-directional. The Dutch text is aligned to the English text and vice versa (dotted lines versus continuous lines). The alignment models produced are asymmetric. Several heuristics exist to combine directional word alignments in order to produce symmetric alignments. These techniques are usually called symmetrisation heuristics and various variants exist. Common techniques start with the intersection of both link sets and iteratively add additional links from the union of both sets. Several constraints can be set to restrict the number of selected links in order to balance precision and recall. Links can be seen in a two-dimensional space with words from the source language in one dimension and words from the target language in the other. Word-to-word links correspond in this setting to points in that space (see the illustration in Figure 2).

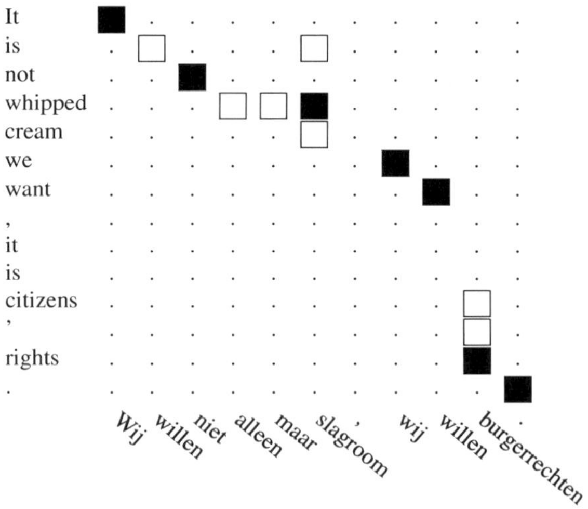

Fig. 2 A word alignment matrix. The intersection of links is illustrated by filled boxes. Additional alignment points from the union of links are shown as empty boxes.

A common symmetrisation strategy is then to add points (links) in this alignment space, which are connected to existing ones. Starting with the intersection, new links can be added iteratively from the union of all links, until no other connected link can be found. This is commonly called the *grow* heuristics, which can be extended by allowing diagonal connections and by extending the final set with links between elements that have so far not been linked to anything at all. Finally, symmetric alignments between words enable the identification of phrase correspondences using simple phrase extraction strategies. More details about phrase extraction can be found in Section 3.2.

2.2 Measures for Computing Semantic Similarity

Translational co-occurrence vectors are used to find distributionally similar words. For simplicity an example of a single-word term *kat* in Table 1 is given. In the current setting the terms can be both single- or multiword terms. Every cell in the vector refers to a particular translational co-occurrence type.[2] For example, *kat* 'cat' gets the translation *Katze* in German. The value of these cells indicate the number of times the co-occurrence type under consideration is found in the corpus. For ease of reading this example gives translations for the single word *kat*, but the alignments are by no means restricted to single words. Instead of *kat* the term *werkzame stof* 'active ingredient' could be used. This is why the remainder of this chapter refers to head terms instead of head words.

Each co-occurrence type has a cell frequency. Likewise each head term has a row frequency. The row frequency of a certain head term is the sum of all its cell frequencies. In the example the row frequency for the term *kat* 'cat' is 65. Cut-offs for cell and row frequency can be applied to discard certain infrequent co-occurrence types or head terms respectively. There is little confidence in characterisations of terms with low frequency. For example, the English translation 'the' in Table 1 has a frequency of 1. A cut-off of 2 would discard this co-occurrence.

Table 1 Translational co-occurrence vector for *kat* based on four languages.

	Katze-DE	chat-FR	gatto-IT	cat-EN	the-EN	*Total*
kat 'cat'	17	26	8	13	1	65

The more similar the vectors are, the more distributionally similar the head terms are. A way to compare the vectors is needed for any two head terms to be able to express the similarity between them by means of a score. Various methods can be used to compute the distributional similarity between terms. In Section 3.4.2 what measures were chosen in the current experiments will be explained.

[2] Language abbreviations are taken from the ISO-639 2-letter codes.

Methods for computing distributional similarity between two words often consist of a measure for computing the similarity between two co-occurrence vectors and a measure for assigning weights to the co-occurrence types present in the vector. Feature weights have been used in previous work for syntax-based methods to account for the fact that co-occurrences have different information values. Selectionally weak (Resnik, 1993) or *light* verbs such as *hebben* 'to have' have a lower information value than a verb such as *uitpersen* 'squeeze' that occurs less frequently. Although weights that promote features with a higher information value work very well for syntax-based methods, van der Plas (2008b) showed that weighting only helps to get better synonyms for very infrequent nouns when applied in alignment-based approaches. In the current setting very infrequent terms are not considered so no weightings were used.

2.3 Related Work

Multilingual parallel corpora have mostly been used for tasks related to word sense disambiguation, such as target word selection (Dagan et al, 1991) and separation of senses (Resnik and Yarowsky, 1997; Dyvik, 1998; Ide et al, 2002).

However, taking sense separation as a basis, Dyvik (2002) derives relations such as synonymy and hyponymy by applying the method of semantic mirrors. The paper illustrates how the method works. Firstly, different senses are identified on the basis of manual word translations in sentence-aligned Norwegian-English data (2,6 million words in total). Secondly, senses are grouped in semantic fields. Thirdly, features are assigned on the basis of inheritance. Lastly, semantic relations such synonymy and hyponymy are detected based on interrelationships and inclusion among feature sets .

The following two papers are driven by the same motivation, namely, to improve the syntax-based methods that are not precise enough to find synonyms. However, both papers discussed below have taken bilingual dictionaries as a starting point and not corpora and they are limited to single-word terms.

Lin et al (2003) try to tackle the problem of identifying synonyms in lists of nearest neighbours in two ways: Firstly, they look at the overlap in translations of semantically similar words in multiple bilingual dictionaries. Secondly, they design specific patterns designed to filter out antonyms. They evaluate a set of 80 synonyms and 80 antonyms from a thesaurus, which are also found among the top-50 words that are distributionally similar to each other. The pattern-based method results in a precision of 86.4% and a recall of 95.0%. The method using bilingual dictionaries gets a higher precision score (93.9%). However, recall is much lower: 39.2%.

Wu and Zhou (2003) report an experiment on synonym extraction using bilingual resources (an English-Chinese dictionary and corpus) as well as monolingual resources (an English dictionary and corpus). Their monolingual corpus-based approach is very similar to the monolingual corpus-based approach. The bilingual approach is different from ours in several aspects. Firstly, they do not take the corpus as the starting point to retrieve word alignments. They use the bilingual dictionary to retrieve multiple translations for each target word. The corpus is only employed to assign probabilities to the translations found in the dictionary. The authors praise the method for being able to find synonyms that are not in the corpus as long as they are found in the dictionary. However, the drawback is that the synonyms are limited to the coverage of the dictionary. The aim of automatic methods in general is precisely to overcome the limited coverage of such resources. A second difference with the system is the use of a bilingual parallel corpus whereas a multilingual corpus is used containing 11 languages in total. The authors show that the bilingual method outperforms the monolingual methods both in recall and precision. However, a combination of different methods leads to the best performance. A precision of 27.1% on middle-frequency nouns is attained.

Van der Plas and Tiedemann (2006) used a distributional method to find single-word synonyms in parallel corpora for the general domain. The method is very similar to the methods proposed in this chapter, apart from the fact that in van der Plas and Tiedemann (2006) intersective single-word alignments are used instead of phrases, and the method is applied to texts from the general domain.

Some researchers present methods for the automatic acquisition of paraphrases, including multi- and single-word synonyms (Barzilay and McKeown, 2001; Ibrahim et al, 2003; Shimota and Sumita, 2002; Bannard and Callison-Burch, 2005; Callison-Burch, 2008). The first two of these have used a monolingual parallel corpus to identify paraphrases. The last three employ multilingual corpora.

Barzilay and McKeown (2001) present an unsupervised learning approach for finding paraphrases from a corpus of multiple English translations of the same source text. They trained a classifier with the help of identical words (co-training). The method retrieved 9,483 lexical paraphrases of which 500 were selected for evaluation. 70.8% were single words. A manual evaluation resulted in an average precision of 85.5%. Evaluation on WordNet only resulted in 35% of the paraphrases being synonyms. 32% are in a hypernym relation, 18% are siblings and 10% are unrelated. Examples of paraphrases classified as hypernyms by WordNet are *landlady* and *hostess* and *reply* and *say*. Examples of siblings (co-hyponyms) are *city* and *town*, and *pine* and *fir*. The authors argue that synonymy is not the only source of paraphrasing. It is a fact that people can paraphrase by using alternative wording that can be more specific or more general in nature than the original wording. Although hypernyms and hyponyms of the original words are used in the second phrase, people do still judge it to be a paraphrase.

Ibrahim et al (2003) present an approach that is a synthesis of Barzilay and McKeown (2001) and a method based on dependency tree paths between a pair of words in monolingual data by Lin and Pantel (2001). Ibrahim et al (2003) capture long-distance dependencies with structural paraphrases, generalising syntactic paths. This way they hope to find longer paraphrases. Indeed the average length of the paraphrases learned reaches 3.26. The precision of 130 paraphrases according to three human judges is on average 41,2%.

Shimota and Sumita (2002) propose a method for extracting paraphrases from a bilingual corpus of approximately 162K sentences of travel conversation. They select all sentences with the same translations. Extraction and filtering is done on the basis of Dynamic Programming (DP) matching (Cormen et al, 2001). Only sentences that differ in fewer than 4 words are selected. Variant words and surrounding words are extracted. At last filtering is done on the basis of frequency and association strength. A manual evaluation was carried out in which judges had to label a candidate paraphrase as *same*, *different*, *semantically improper*, *syntactically improper*. 83.1% of the candidate paraphrases for the English-Japanese setting were labelled as being similar. 93.5% of the candidate paraphrases for the Japanese-English setting were considered the same.

Bannard and Callison-Burch (2005) use a method that is rooted in phrase-based statistical machine translation. Translation probabilities provide a ranking of candidate paraphrases. These are refined by taking contextual information into account in the form of a language model. The Europarl corpus (Koehn, 2005) is used. It has about 30 million words per language. 46 English phrases are selected as a test set for manual evaluation by two judges. When using automatic alignment, the precision reached without using contextual refinement is 48.9%. A precision of 55.3% is reached when using context information. Manual alignment improves the performance by 26%. A precision score of 55% is attained when using multilingual data.

In a more recent publication Callison-Burch (2008) improved the method in Bannard and Callison-Burch (2005) by using syntactic constraints and multiple languages in parallel. This study implemented a combination of Bannard and Callison-Burch (2005) and Callison-Burch (2008), in which the results were compared with POS filters, instead of syntactic constraints. More details can be found in the Section 5.

In addition to methods that use parallel corpora, mono-lingual pattern-based methods have also been used to find term variation. One example of this type of study has taken place within the framework of the IMIX project. Fahmi (2009) acquired term variation for the medical domain using a two-step model. As a first step an initial list of synonyms are extracted using a method adapted from DIPRE (Brin, 99). During this step syntactic patterns guide the extraction of candidate terms in the same way as they will guide the extraction in this chapter. This first step results in a list of candidate synonyms that are further filtered following a method described in Lin et al (2003), which uses Web pages as an external source to measure the synonym compatibility hits of each pair. The precision and recall scores presented in Fahmi (2009) are high. The test set in Section 5 will give results for this method and refer to it as the pattern-and web-based approach.

3 Materials and Methods

In the following subsections the set up for the experiments is described. After describing the corpora (3.1) and the translations extracted from them (3.3) they were compared with the resulting vectors in Subsection 3.4.2. The section finishes with a description of some additional post-processing that was done (3.5).

3.1 The multilingual Parallel Corpus EMEA

Measures of distributional similarity usually require large amounts of data. For the alignment method a parallel corpus of reasonable size is needed with Dutch either as source or as target language. Furthermore, their is a wish to experiment with various languages aligned to Dutch.

The freely available EMEA corpus (Tiedemann, 2009) includes 22 languages in parallel with a reasonable size of about 12-14 million tokens per language. The entire corpus is aligned at the sentence level for all possible combinations of languages. Thus, for acquiring Dutch synonyms there are 21 language pairs with Dutch as the source language: Bulgarian (BG), Czech (CS), Danish (DA), German (DE), Greek (EL), English (EN), Spanish (ES), Estonian (ET), Finnish (FI), French (FR), Hungarian (HU), Italian (IT), Latvian (LT), Lithuanian (LV), Maltese (MT), Polish (PL), Portuguese (PT), Romanian (RO), Slovak (SK), Slovene (SL), and Swedish (SV). Each language pair includes about 1.1 million sentence pairs. Note that there is a lot of repetition in EMEA and the number of unique sentences is much smaller: around 350,000 sentence pairs per language pair with about 6-7 million tokens per language. Parallel translations of all languages were used to extract Dutch term variants.

3.2 Automatic Word Alignment and Phrase Extraction

For sentence alignment,*hunalign* (Varga et al, 2005) was applied, with the 'realign' function that induces the combination of lexical features from the bitext with length-based features. Word alignment was performed using the well-known IBM alignment models (Brown et al, 1993) implemented in the open-source tool GIZA++ (Och, 2003). Standard settings, defined in the Moses tool kit for statistical machine translation (Koehn et al, 2007), were used to generate Viterbi word alignments of IBM model 4, for sentences not longer than 80 tokens. In order to improve the statistical alignment, lower-case tokens and, where available, lemmas (produced by the *Tree-Tagger* (Schmid, 1994) and the Alpino parser (van Noord, 2006)) were used.

The *grow* heuristics are used to combine the asymmetric word alignments that starts with the intersection of the two Viterbi alignments and adds block-neighbouring points to it in a second step (as explained in Section 2.1). The way high precision links were obtained with some many-to-many alignments. Finally the phrase extraction tool was used from Moses to extract phrase correspondences. Phrases in statistical machine translation are defined as sequences of consecutive words and phrase extraction, which refers to the extraction of all possible phrase pairs that are consistent with the underlying word alignment. Consistency in this case means that words in a legal phrase are only aligned to words in the corresponding phrase, and not to any other word outside that phrase. The extraction mechanism can be restricted by setting a maximum phrase length, which is seven in the default settings of Moses. In the experiments, the maximum phrase length was set to 4, because terms in the medical domain are not expected to exceed 4 words.

As explained above, word alignment is carried out on lower-case and possibly lemmatised versions of the corpus. However, for phrase extraction, surface word forms were used and were extracted, along with the part-of-speech tags for Dutch, taken from the Alpino parse trees. This allowed all lower-case words except those with *name* as part-of-speech. Furthermore, it allowed filtering of the resulting phrase table according to POS patterns defined for the extraction of multiword terms. Phrases consisting of only non-alphabetical characters were also removed.

3.3 Selecting Candidate Terms

Those phrases that are more likely to be good terms can be used by using a regular expression that describes the sequence of POS tags of possible (or candidate) terms. The regular expression applied in this study describes a pattern of adjectives (A), nouns (NN), names (NM) and prepositions (P) and was adapted to Dutch by Fahmi et al (2007) from Justeson and Katz. (1995):

```
((A|NN|NM)+|(((A|NN|NM)*(NN|NM P)?)(A|NN|NM)*))NN+
```

To explain this regular expression in words, a candidate term is either a sequence of adjectives and/or nouns and/or names, ending in a noun or name or it consists of two such strings, separated by a single preposition. Some examples of permitted patterns are *werkzame/A stof/NN* ‘active substance’, *symptomen/NN van/P agitatie/NN* ‘symptoms of agitation’ , and *werkzaamheid/NN van/P Abilify/NM* ‘efficacy of Abilify’.

After applying the filters and removing all hapaxes, the 9.76 M co-occurrences of a Dutch (multiword) term and a foreign translation are left.

Note that data relies entirely on automatic processing. Thus, results from the automatic word alignments include errors. Bannard and Callison-Burch (2005) show that when using manual alignment the percentage of correct paraphrases rises from 48.9% to 74.9%. It is clear that the automatic alignment introduces a lot of noise.

3.4 Comparing Translation Vectors

To compare the vectors of the terms it is a similarity of measures which is needed. Definitions and description of the similarity measure chosen to be used in these experiments were given.

3.4.1 Definitions

The functions used in this chapter are described using an extension of the notation used by Lin (1998), adapted by Curran (2003). Co-occurrence data is described as tuples: $\langle$word, language, word$'\rangle$, for example, $\langle$kat, EN, cat$\rangle$.

Asterisks indicate a set of values, ranging over all existing values of that component of the relation tuple. For example, $(w,*,*)$ denotes for a given word w all translational contexts it has been found in, in any language. For the example of *kat*, this would denote all values for all translational contexts the word is found in: *Katze_DE*:17, *chat_FR*:26 etc. Everything is defined in terms of co-occurrence data with non-zero frequencies.

The set of attributes or features for a given corpus is defined as:

$$(w,*,*) \equiv \{(r,w')|\exists(w,r,w')\}$$

Each pair yields a frequency value, and the sequence of values is a vector indexed by r:w' values, rather than natural numbers. A subscripted asterisk indicates that the variables are bound together:

$$\sum(w_m,*_r,*_{w'}) \times (w_n,*_r,*_{w'})$$

The above refers to a dot product of the vectors, for term w_m and term w_n, summing over all the r:w' pairs these two terms have in common. For example, the vectors for *kat* could be compared with some other terms by applying the dot product to all bound variables.

3.4.2 Similarity Measure

The experiments are limited to using Cosine. These methods were chosen, since they performed best in experiments reported in van der Plas (2008a).

Cosine is a geometrical measure. It returns the cosine of the angle between the vectors of the words and is calculated as the dot product of the vectors:

$$Cosine = \frac{\sum(W1,*_r,*_{w'}) \times (W2,*_r,*_{w'})}{\sqrt{\sum(W1,*,*)^2 \times \sum(W2,*,*)^2}}$$

If the two words have the same distribution, the angle between the vectors is zero. The maximum value of the Cosine measure is 1.

3.5 Post-processing

A well-known problem of phrase-based methods to paraphrase or term variation acquisition is the fact that a large proportion of the term variants or paraphrases proposed by the system are super- or sub-strings of the original term (Callison-Burch, 2008).[3] To remedy this problem all term variants were removed that are either super- or sub-strings of the original term from the lists of candidate term variants output by the system. For example, the system outputs many super strings of the term *zuurstof* 'oxygen' that are not term variants, such as *zuurstoftekort* 'lack of oxygen' *zuurstofverbruik* 'usage of oxygen'. This post-processing removes a large proportion of the term variants proposed by the system. When looking at the top-three candidate term variants around 30% of the terms are either super- or sub-strings.

4 Evaluation

Evaluation of terminology extractors is not a trivial task (Vivaldi and Rodríguez, 2007). This chapter evaluates both term extraction and term variation extraction at the same time. There are several evaluation methods available to assess lexico-semantic data. Curran (2003) distinguishes two types of evaluation: direct evaluation and indirect evaluation. Direct evaluation methods compare the semantic relations given by the system against human performance or expertise. Indirect approaches do not use human evidence directly, the system is evaluated by measuring its performance on a specific task. Such approaches are referred to as task-based evaluation. The direct approaches can be subdivided in comparisons against gold standards (for example, EWN, synonym lists, association lists) and comparisons against ad hoc human judgements, i.e. manual evaluations of the output of the system.

Many NLP tasks can be evaluated using a gold standard. In parsing for example one might compare the results of the system with the ones provided in a manually annotated tree bank. In previous work on synonym acquisition for the general domain van der Plas and Tiedemann (2006) used the synsets in Dutch EuroWordnet (Vossen, 1998) for the evaluation of the proposed synonyms. In EWN one synset consists of several synonyms which represent a single sense. Polysemous words occur in several synsets. However, the gold standard used proved to be incomplete. In an evaluation with human judgements van der Plas and Tiedemann (2006) showed that in 37% of the cases the majority of the subjects judged the synonyms proposed by the system to be correct even though they were not found to be synonyms in Dutch EuroWordnet.

---

[3] This is the case for single-word alignment-based approaches as well, as discussed in van der Plas (2008b).

An example of a gold standard for the medical domain would be the UMLS resource, described earlier, a large database in the domain of health and biomedicine for the English language. For evaluating medical term variation in Dutch there are not many gold standards available. Moreover, the gold standards that are available are incomplete.

4.1 Gold Standard

The term variants proposed by the systems under consideration on the list of term variants from Elseviers medical encyclopedia: a medical encyclopedia intended for the general audience containing 379K words wer chosen for evaluation. The encyclopedia was made available by Spectrum b.v., and can also be found on-line.[4]

4.2 Test Set

The test set comprises of 848 medical terms from *aambeeld* 'incus' to *zwezerik* 'thymus' and their term variants. Multiword terms are found in 258 of these entries. For most of the terms the list from Elseviers medical encyclopedia gives only one term variation, 146 terms have two term variants and only one term has three term variants. For each of these medical terms in the test set the system generates a ranked list of term variants that will be evaluated against the term variants in the gold standard.

5 Results and Discussion

Before presenting results and giving some error analysis the reader is to be reminded of the two methods the results were compared with and give some more detail on the implementation of the second method.

5.1 Two Methods for Comparison

The first method is the pattern- and web-based approach described in Fahmi (2009). Note that the method was not re-implemented, so the method was not able to run on the same corpus in the experiments being used. The corpus used in Fahmi (2009) is the IMIX medical corpus developed in Tilburg University as part of the

[4] `http://www.kiesbeter.nl/medischeinformatie/`

IMIX/Rolaquad project.[5] It consists of texts from a medical encyclopedia and a medical handbook and contains 57,004 sentences. The system outputs a ranked list of term variation pairs. The top-100 pairs were selected that are output by the system and these were evaluated on the test set described in Subsection 4.2. The method is composed of two main steps. In the first steps candidate terms are extracted from the corpus using a POS filter, that is similar to the POS filter applied. In the second step pairs of candidate term variants are re-ranked on the basis of information from the Web. Phrasal patterns such as P_{or} are used to get synonym compatibility hits as opposed to P_{and} that points to non-synonymous terms.

The second method to be compared with is the phrase-based translation method first introduced by Bannard and Callison-Burch (2005). Statistical word alignment can be used to measure the relation between source language items. Here, the estimated translation likelihoods of phrases ($p(f|e)$ and $p(e|f)$) make use of the build translation models in standard phrase-based statistical machine translation systems (Koehn et al, 2007). Bannard and Callison-Burch (2005) define the problem of paraphrasing as the following search problem:

$$\hat{e_2} = \underset{e_2:e_2 \neq e_1}{\operatorname{argmax}}\, p(e_2|e_1)$$

where

$$p(e_2|e_1) = \sum_f p(f|e_1)p(e_2|f,e_1) \approx \sum_f p(f|e_1)p(e_2|f)$$

For paraphrasing the analysis is not only interested in $\hat{e_2}$, but also in the top-ranked phrase candidates. However, this does not change the algorithm. In their paper, they also show that systematic errors (usually originating from bad word alignments) can be reduced by summing over several language pairs.

$$p(e_2|e_1) \approx \sum_C \sum_{f_C} p(f_C|e_1)p(e_2|f_C)$$

This is the approach that was also adapted for comparison. The only difference in the implementation is that a POS-filter is applied to extract candidate terms as explained in section 3.3. In some sense this is a sort of syntactic constraint as one of the authors introduces in a later publication (Callison-Burch, 2008). The maximum phrase length is set to 4 and applied the same post-processing as described in Subsection 3.5.

5.2 Results

Table 2 shows the results for the method compared with the method adapted from Bannard and Callison-Burch (2005) and Callison-Burch (2008) and the method by

---

[5] `http://ilk.uvt.nl/rolaquad`

Table 2 Percent precision and recall at several values of k and percent coverage for the method proposed in this chapter, the method adapted from Bannard and Callison-Burch (2005) and the output of the system proposed by Fahmi (2009).

Method	k=1		k=2		k=3		Coverage
	P	R	P	R	P	R	
Phrase-based Distributional Similarity	28.9	22.8	21.8	32.7	17.3	37.2	40.0
Bannard & Callison-Burch (2005)	20.9	17.5	16.8	27.2	13.4	31.7	47.7
Fahmi (2009)	38.2	35.1	37.1	35.1	37.1	35.1	4.0

Fahmi (2009). Precision and recall are given at several values of k. At k=1, only the top-1 term variants the system proposes are taken into account. At k=3 the top-3 candidate term variants are included in the calculations.

The last column shows the coverage of the system. A coverage of 40% means that from 340 of the 850 terms in the test set, one or more term variants are found.

From Table 2 one can read that the method proposed is able to get about one third of the term variants right, when only the top-1 candidates are considered. It is able to retrieve roughly a quarter of the term variants provided in the gold standard[6]. If k increased precision goes down and recall goes up. This is expected, because the system proposes a ranked list of candidate term variants so at higher values of k the quality is lower, but more terms from the gold standard are found.

In comparison, the scores resulting from the adapted implementation of Bannard and Callison-Burch (2005) are lower. They do however, manage to find more terms from the test set covering 47.7% of the words in the gold standard. This is due to the cut-off that is used when creating the co-occurrence vector. In the approach hapaxes were discarded to obtain the scores in Table 2 whereas for the Bannard & Callison-Burch approach the entire phrase table is used. This brings up the question whether the differences in performance are simply due to this cut-off that removes unreliable data points. Therefore the system was again rerun without this cut-off. As expected, the coverage went up in that setting – actually to 47.7% as well. However, the precision and recall remained higher, than the scores with the implementation following Bannard and Callison-Burch (2005): 25.4% and 20.9%, respectively. The vector-based approach seems to outperform the direct use of probabilities from phrase-based MT.

Finally, the results were also compared with the data set extracted using the pattern- and web-based approach from Fahmi (2009). The precision and recall figures of that data set are the highest in the comparison. However, since the coverage of this method is very low, which is not surprising since a smaller corpus is used to get these results, the precision and recall are calculated on the basis of a very small number of examples (35 to be precise). The results are therefore not very reliable. The precision and recall figures presented in Fahmi (2009), however, point in the same direction. To get an idea of the actual coverage of this method this extraction technique would need to be applied to the EMEA corpus. This is

[6] Note that a recall of 100% is not possible, because some terms have several term variants.

especially difficult due to the heavy use of web queries which makes it problematic to apply it to large data sets.

5.3 Error Analysis

100 candidate term variants not found in the gold standard were manually inspected to get an idea of the types of mistakes the system makes.

The most important finding was that many of the term variants proposed by the system (25%) that are not found in the gold standard, are in fact correct. This number is just an approximation. A manual evaluation with a domain specialist would give more realistic, but probably still much higher scores than the scores given in Table 2. Here, are some examples below:

aangezichtsverlamming	gelaatsparalyse	'facial paralysis'
ademnood	ademhalingsnood	'respiratory distress'
alvleesklierkanker	pancreaskanker	'cancer of the pancreas'
cervix	baarmoederhals	'cervix'

Some mistakes could have been avoided using stemming or proper lemmatisation (plurals that are counted as wrong). The introduction explained that this was not concerned with inflectional variation, because it works on normalised forms. Unfortunately the lemmatisation and stemming is not flawless. Approximately around 8% of the errors are due to mistakes in stemming:

abortus	'abortion'	zwangerschapsafbrekingen	'abortions'
adenoom	'adenoma'	adenomen	'adenomas'
indigestie	'indigestion'	spijsverteringsstoornissen	'indigestion problems'

Another major source of error is translations of the headword that are proposed by the system as term variants (9%). These are the result of non-translated text in the Dutch part of the corpus. In this task these translations of terms into multiple languages are counted as wrong, but it can well be imagined that the multilingual approach proposed here offers possibilities for multilingual term variation extraction.

astma	asthma	'asthma'
geslacht	gender	'gender'

Some spelling variants (2%) were also found which are usually not covered by the gold standard. Look, for instance, at the following examples:

diarree diaree 'diarrhea'
faeces feces 'faeces'

After removing the previous cases from the data almost half as many incorrect term variants are left, some of which are related to the problem mentioned in Section 3.5: Phrase-based methods to paraphrase or term variation acquisition have the tendency to propose term variants that are super- or sub-strings of the original term (Callison-Burch, 2008). These super- or sub-strings were able to be filtered out, but not in cases where a candidate term is a term variation of a super- or sub-string of the original term. Consider, for example the term *bloeddrukverlaging* 'blood pressure decrease' and the candidate *afname* 'decrease', where *afname* is a synonym for *verlaging*.

6 Conclusions

Measures of distributional similarity using translational context have been shown to be effective for the identification of medical term variation. Standard techniques for automatic word alignment and phrase extraction coming from research on statistical machine translation can be employed to collect translational variation across various languages. This type of variation is then used to measure semantic similarity between words and phrases. This technique is compared with a pattern-based filter using part-of-speech labels to focus on particular constructions which are common among multiword terms. In the experiments with Dutch terms and a parallel corpus of 22 European languages with texts from the medical domain the methods are shown to outperform another alignment-based approach measured on a gold standard taken from a medical encyclopedia. Precision and recall are still quite poor according to the automatic evaluation. However, manual inspection suggests that many candidates are simply misjudged because of the low coverage of the gold standard data. In conclusion the approach is thought to provide a promising strategy for extracting term variants using fully-automatic techniques. In the research the main focus was set to build resources for improving the coverage of a question-answering system. However, it is thought that the outcome of such extraction runs can be useful for a wide range of applications and for the extension of existing lexical resources and ontologies. Especially encouraging is that the general technique is robust and flexible enough to be applied to various languages and domains ans this has already been shown in related studies. Moreover, due to the multilingual nature of the method it has great potential for the extraction of multilingual term variation.

Acknowledgements

This research has mainly been carried out in the project *Question Answering using Dependency Relations*, which is part of the research programme for *Interactive Multimedia Information eXtraction*, IMIX, financed by NWO, the Dutch Organisation for Scientific Research. Part of this work has received funding from the EU FP7 programme (FP7/2007-2013) under grant agreement nr 216594 (CLASSIC project: `www.classic-project.org`).

References

Bannard C, Callison-Burch C (2005) Paraphrasing with bilingual parallel corpora. In: Proceedings of the annual Meeting of the Association for Computational Linguistics (ACL)

Barzilay R, McKeown K (2001) Extracting paraphrases from a parallel corpus. In: Proceedings of the Annual Meeting of the Association for Computational Linguistics (ACL), pp 50–57, URL `citeseer.ist.psu.edu/barzilay01extracting.html`

Bouma G, Fahmi I, Mur J, van Noord G, van der Plas L, Tiedemann J (2007) Linguistic knowledge and question answering. Traitement Automatique des Langues (TAL) 2005(03)

Brin S (99) Extracting patterns and relations from the World Wide Web. In: WebDB '98: Selected papers from the International Workshop on The World Wide Web and Databases

Brown P, Della Pietra S, Della Pietra V, Mercer R (1993) The mathematics of statistical machine translation: Parameter estimation. Computational Linguistics 19(2):263–296

Callison-Burch C (2008) Syntactic constraints on paraphrases extracted from parallel corpora. In: Proceedings of EMNLP

Cormen T, Leierson C, Rivest R, Stein C (2001) Introduction to algorithms. MIT Press

Curran J (2003) From distributional to semantic similarity. PhD thesis, University of Edinburgh

Dagan I, Itai A, Schwall U (1991) Two languages are more informative than one. In: Proceedings of the Annual Meeting of the Association for Computational Linguistics (ACL)

Dyvik H (1998) Translations as semantic mirrors. In: Proceedings of Workshop Multilinguality in the Lexicon II (ECAI)

Dyvik H (2002) Translations as semantic mirrors: from parallel corpus to wordnet. Language and Computers, Advances in Corpus Linguistics Papers from the 23rd International Conference on English Language Research on Computerized Corpora (ICAME 23) 16:311–326

Fahmi I (2009) Automatic term and relation extraction for medical question answering system. PhD thesis, University of Groningen

Fahmi I, Bouma G, van der Plas L (2007) Using multilingual terms for biomedical term extraction. In: Proceedings of the RANLP Workshop on Acquisition and Management of Multilingual Lexicons , Borovetz, Bulgaria

Fellbaum C (1998) WordNet, an electronic lexical database. MIT Press

Firth J (1957) A synopsis of linguistic theory 1930-1955. Studies in Linguistic Analysis (special volume of the Philological Society) pp 1–32

Furnas G, Landauer T, Gomez L, Dumais S (1987) The vocabulary problem in human-system communication. In: Communications of the ACM, pp 964–971

Harris Z (1968) Mathematical structures of language. Wiley

Ibrahim A, Katz B, Lin J (2003) Extracting structural paraphrases from aligned monolingual corpora. In: Proceedings of the second international workshop on Paraphrasing (IWP), pp 57–64

Ide N, Erjavec T, Tufis D (2002) Sense discrimination with parallel corpora. In: Proceedings of the ACL Workshop on Sense Disambiguation: Recent Successes and Future Directions

Justeson J, Katz S (1995) Technical terminology: some linguistic properties and an algorithm for identification in text. Natural Language Engineering 1:9–27

Kilgarriff A, Yallop C (2000) What's in a thesaurus? In: Proceedings of the Second Conference on Language Resource an Evaluation (LREC)

Koehn P (2005) Europarl: A parallel corpus for statistical machine translation. In: Proceedings of the MT Summit, Phuket, Thailand, pp 79–86

Koehn P, Hoang H, Birch A, Callison-Burch C, MFederico, Bertoldi N, Cowan B, Shen W, Moran C, Zens R, Dyer C, Bojar O, AConstantin, Herbst E (2007) Moses: Open source toolkit for statistical machine translation. In: Proceedings of the Annual Meeting of the Association for Computational Linguistics

Lin D (1998) Automatic retrieval and clustering of similar words. In: Proceedings of COLING/ACL

Lin D, Pantel P (2001) Discovery of inference rules for question answering. Natural Language Engineering 7(4):343-360 7(4):343–360

Lin D, Zhao S, Qin L, Zhou M (2003) Identifying synonyms among distributionally similar words. In: Proceedings of the International Joint Conference on Artificial Intelligence (IJCAI)

McCray A, Hole W (1990) The scope and structure of the first version of the umls semantic network. In: Symposium on Computer Applications in Primary Care (SCAMC-90), , Washington DC, IEEE Computer Society. 126-130., IEEE Computer Society, pp 126–130

van Noord G (2006) At last parsing is now operational. In: Actes de la 13eme Conference sur le Traitement Automatique des Langues Naturelles

Och F (2003) GIZA++: Training of statistical translation models. Available from `http://www.isi.edu/~och/GIZA++.html`

van der Plas L (2008a) Automatic lexico-semantic acquisition for question answering. Groningen dissertations in linguistics van der Plas L (2008b) Automatic lexico-semantic acquisition for question answering. PhD thesis, University of Groningen

van der Plas L, Tiedemann J (2006) Finding synonyms using automatic word alignment and measures of distributional similarity. In: Proceedings of COLING/ACL

van der Plas L, Tiedemann J (2010) Finding medical term variations using parallel corpora and distributional similarity. In: Proceedings of the Coling workshop on ontologies and lexical resources

Resnik P (1993) Selection and information, unpublished doctoral thesis, University of Pennsylvania

Resnik P, Yarowsky D (1997) A perspective on word sense disambiguation methods and their evaluation. In: Proceedings of ACL SIGLEX Workshop on Tagging Text with Lexical Semantics: Why, what, and how?

Roget P (1911) Thesaurus of English words and phrases

Schmid H (1994) Probabilistic part-of-speech tagging using decision trees. In: Proceedings of International Conference on New Methods in Language Processing, Manchester, UK, pp 44–49, http://www.ims.uni-stuttgart.de/˜schmid/

Schütze H (1992) Dimensions of meaning. In: Proceedings of the ACM/IEEE conference on Supercomputing

Shimota M, Sumita E (2002) Automatic paraphrasing based on parallel corpus for normalization. In: Proceedings of the International Conference on Language Resources and Evaluation (LREC)

Tiedemann J (2009) News from OPUS - A collection of multilingual parallel corpora with tools and interfaces. In: Nicolov N, Bontcheva K, Angelova G, Mitkov R (eds) Recent Advances in Natural Language Processing, John Benjamins, Amsterdam/Philadelphia, Borovets, Bulgaria, vol V, pp 237–248

Varga D, Németh L, Halácsy P, Kornai A, Trón V, Nagy V (2005) Parallel corpora for medium density languages. In: Proceedings of RANLP 2005, pp 590–596

Ville-Ometz F, Royauté J, Zasadzinski A (2008) Enhancing in automatic recognition and extraction of term variants with linguistic features. Terminology, International Journal of Theoretical and Applied Issues in Specialized Communication 13:1:35–59

Vivaldi J, Rodríguez H (2007) Evaluation of terms and term extraction systems: A practical approach. Terminology, International Journal of Theoretical and Applied Issues in Specialized Communication 13:2:225–248

Vossen P (1998) EuroWordNet a multilingual database with lexical semantic networks

Wilks Y, Fass D, Guo CM, McDonald JE, T Plate BMS (1993) Providing machine tractable dictionary tools. Machine Translation 5(2):99–154

Wu H, Zhou M (2003) Optimizing synonym extraction using monolingual and bilingual resources. In: Proceedings of the International Workshop on Paraphrasing: Paraphrase Acquisition and Applications (IWP)

Relation Extraction for Open and Closed Domain Question Answering

Gosse Bouma, Ismail Fahmi and Jori Mur

Abstract One of the most accurate methods in Question Answering (QA) uses off-line information extraction to find answers for frequently asked questions. It requires automatic extraction from text of all relation instances for relations that users frequently ask for. In this chapter, two methods are presented for learning relation instances for relations relevant in a closed and open domain (medical) QA system. Both methods try to learn automatic dependency paths that typically connect two arguments of a given relation. The first (lightly supervised) method starts from a seed list of argument instances, and extracts dependency paths from all sentences in which a seed pair occurs. This method works well for large text collections and for seeds which are easily identified, such as named entities, and is well-suited for open domain QA. A second experiment concentrates on medical relation extraction for the question answering module of the IMIX system. The IMIX corpus is relatively small and relation instances may contain complex noun phrases that do not occur frequently in the exact same form in the corpus. In this case, learning from annotated data is necessary. Dependency patterns enriched with semantic concept labels are shown to give accurate results for relations that are relevant for a medical QA system. Both methods improve the performance of the Dutch QA system *Joost*.

Gosse Bouma
University of Groningen, Groningen, The Netherlands, e-mail: `g.bouma@rug.nl`

Ismail Fahmi
Gresnews Media, Amsterdam, The Netherlands, e-mail: `ismail.fahmi@gmail.com`

Jori Mur
De Rode Planeet, Zuidhorn, The Netherlands, e-mail: `jori.mur@gmail.com`

1 Introduction

QA is the task of finding answers to user questions in (large) text collections. Most QA systems use a technique whereby questions are first analysed to determine the expected answer type. Passage retrieval is subsequently used to retrieve the most relevant text fragments from the text collection. Various NLP techniques then help identify and rank phrases within the retrieved text that potentially answer the question and match the expected answer type. An alternative method searches the corpus beforehand for answers to frequently asked question types. It requires that relation extraction patterns are constructed (manually or automatically) that extract tuples of pairs instantiating a particular relation from the corpus. For instance, if questions about locations of museums are frequent, one can design patterns that would extract the tuple ⟨*Uffizi, Florence*⟩ from the sentence *Today the Uffizi is one of the most popular tourist attractions of Florence*. For those questions for which relation extraction patterns were developed, the open-domain Dutch QA system *Joost* tends to give better results than using passage retrieval. Bouma et al (2005) show that on questions from CLEF 2003 - 2005, questions answered by consulting tables of previously extracted relation pairs achieve a mean reciprocal rank between 0.74 and 0.88, whereas questions answered by means of passage retrieval and answer extraction from text snippets results in a mean reciprocal rank between 0.53 and 0.62. This suggests that in general it is worthwhile to invest in relation extraction for QA.

Much research on automatic learning of relation extraction patterns has concentrated on learning surface strings (i.e. sequences of words in the immediate context of the arguments of the relation) as extraction patterns. Such patterns can be found in unparsed corpora, and can be used to extract instance pairs from the web. Alternatively, one may learn dependency patterns. Dependency patterns abstract from many aspects of surface word order, and focus on those aspects of grammatical structure that seem most relevant for relation extraction. Below concentrates on the latter approach. For a language such as Dutch, which exhibits more word order variation than English, the fact that dependency patterns abstract over surface word order is important. Another reason for adopting dependency patterns is the fact that the question-answering system *Joost* (Bouma et al, 2005) also works with dependency paths in all other components of the system. Most importantly, for information-retrieval and answer extraction, is that the system relies on the fact that the full text collection in which it can find answers is syntactically parsed. Thus, there is access to dependency information and no additional effort is required.

In this chapter two important issues are addressed relevant to relation extraction for QA. First, manual construction of extraction patterns is labour intensive, especially if many question types need to be dealt with. Thus, it becomes interesting to study the effect of lightly supervised relation extraction methods, that try to learn extraction patterns by bootstrapping from a small set of seed pairs, representative for the relation that needs to be learned. Such methods work well given a large corpus, in which relation instance pairs (such as ⟨*Uffizi, Florence*⟩ for the *museum-location* relation) can be found frequently, and in many different contexts. Below it is shown that, for the purposes of QA, lightly supervised bootstrapping methods can improve the performance of a QA system, even if the accuracy of the automatically retrieved instance pairs is relatively low.

For closed-domain, medical, QA the number of question types is limited. Most questions will be about *definitions, causes, symptoms* and *treatments*. This suggests relation extraction could be very effective for a medical QA system. A problem for domain specific relation extraction, however, is the fact that corpora tend to be smaller than those used for open-domain QA, and thus there are fewer highly frequent instance pairs. Secondly, whereas relation extraction for open domain QA has concentrated on learning relations between named entities, the arguments of medical relations are often complex noun phrases, that are subject to more grammatical variation than named entities. This is an additional factor, that reduces the frequency of easily identifiable instance pairs. Therefore, most systems for relation extraction in the medical domain have made use of two additional resources to make the task feasible. First, a thesaurus (such as UMLS (Bodenreider, 2004)) is used to identify relevant concepts in the text. Second, instead of learning from seeds, extraction patterns are learned, on the basis of an annotated text corpus. In an annotated corpus, examples of the relation to be learned are marked. The relation extraction system then uses these positive examples to learn which grammatical patterns are typical for the relation.

Most work on medical QA has been done for English. For a Dutch medical QA system, relevant resources are less easy to obtain. This section shows that UMLS can also be used for concept labelling Dutch text. Simple string matching, matching of stems, and matching of automatically translated phrases allow a substantial number of Dutch medical terms to be matched with their English counterpart in UMLS. Secondly, relation extraction experiments are performed using the IMIX corpus developed by the IMIX ROLAQUAD team. This is a 60K word corpus of Dutch medical text annotated with semantic concepts and relations. This section shows how a system that extracts instance pairs by means of relation extraction patterns, and filters the results by imposing constraints on the semantic classes to which the arguments must belong, gives accurate results.

In the next section, previous work is discussed on relation extraction for open domain QA and for the medical domain, and this motivated the choice for learning dependency patterns for relation extraction. In Section 3, the QA system *Joost* is briefly introduced and the construction of dependency patterns from parsed corpus data. In Sections 4 and 5 the method for relation extraction in the context of open-domain QA is presented, and for medical QA, respectively. This is concluded by a discussion of the results.

2 Related Work

2.1 Relation Extraction for Open Domain QA

Soubbotin and Soubbotin (2002) presented a QA mechanism which uses predefined surface patterns to extract potential answer phrases. Inspired by the good results of this system in the TREC 2001 evaluation, Ravichandran and Hovy (2002), Fleischman et al (2003) and others investigated techniques for learning such extraction patterns automatically. In particular, these systems find answers to frequently asked question types by bootstrapping from a small list of seed pairs. Each sentence in which the seed appears leads to a potential extraction pattern (consisting, for instance, of the words intervening between the two arguments of the seed pair). The precision of a pattern is estimated by counting how often one finds the correct answer when one of the argument positions is filled in and the resulting string is submitted as a search query for a web search engine. For example, for the birthday relation, the person's name can be supplied, and one can count how often the correct answer is found amongst all the matching strings. Patterns yielding a high precision score are applied to find the answers to questions during the process of QA. Experiments aimed at extracting relations between named entities use the same metric for selecting accurate extraction patterns.

Lita and Carbonell (2004) introduced an unsupervised algorithm that acquires answers off-line while at the same time improving the set of extraction patterns. In their experiments up to 2000 new relations of who-verb types (e.g. *who-invented, who-owns, who-founded etc.*) were extracted from a text collection of several gigabytes starting with only one seed pattern for each type. Etzioni et al (2005) present an overview of their information extraction system KNOWITALL. Extraction patterns are found by instantiating generic rule templates with predicate labels such as *actor* and *city*. Using these patterns a set of seed instances is extracted with which new patterns can be found and evaluated. KNOWITALL introduces the use of a form of point-wise mutual information between the patterns and the extracted terms which is estimated from the Web search engine hit counts to evaluate the extracted facts.

Pantel and Pennacchiotti (2006) describe ESPRESSO, a minimally supervised bootstrapping algorithm that takes as input a few seed instances and iteratively learns surface patterns to extract more instances. In addition to a text corpus of 6.3 million words, the Web is used to increase recall. To evaluate patterns as well as extracted instance pairs, the authors calculate an association score between a pattern and instances based on point-wise mutual information. Section 4 reports briefly on an experiment in which the ESPRESSO method is used to extract relation instance pairs from parsed data.

2.2 Biomedical Relation Extraction

Relation extraction from biomedical text is an active research area. However, this research is restricted almost completely to English (e.g. medline abstracts), and tends to make heavy use of terminological resources such as MESH[1] and UMLS (Unified Medical Language System).[2] Rosario and Hearst (2004) observe, for instance, that "part of the reason for the success of [their relation extraction algorithms] is the use of a large domain-specific lexical hierarchy for generalisation across classes of nouns". Leroy and Chen (2005) stress the importance of concept labelling, by observing that if the name of a gene and a disease co-occur, it is almost certain that there is a (causal) relation between the two.

When trying to apply similar techniques to languages other than English, one immediately runs into the problem that suitable terminological resources are lacking or have only limited coverage. At the same time, attempts at biomedical relation extraction without access to a terminological resource tend to give poor results. Tjong Kim Sang et al (2005), for instance, evaluates the performance of various relation extraction systems for Dutch in the context of a medical question-answering system, and concludes that both recall and precision is low. One of the reasons for low precision is the fact that these systems do not have access to concept labels.

Section 5, concentrates on medical relation extraction for a language other than English. In particular, it addresses the issue of accurate concept labelling of Dutch medical terms and shows that, by combining the Dutch and English parts of UMLS, reasonable coverage and accuracy can be achieved. Braun et al (2005) attempt to do full translations of Dutch medical terms into English on the basis of UMLS, for a cross-lingual information retrieval system, and find that the accuracy of automatic translation is low. The task is different, in that it only needs to assign a semantic concept label to a (Dutch) term, which does not always require a translation that would be useful for IR.

2.3 Using Syntactic Patterns

In contrast to the bootstrapping approaches discussed above patterns based on dependency relations instead of surface patterns are learned. Using dependency relations, the extraction pattern can be simply defined as the shortest path between the two terms in a dependency graph. Surface patterns are typically harder to define in a natural and meaningful way.

Using syntactic (dependency) analysis for relation extraction has become increasingly popular in recent years (Bunescu and Mooney, 2005; Culotta and Sorensen, 2004; Zhao and Grishman, 2005). Most of this work relies heavily on annotated corpora, such as the ACE corpus, in which relations between named

[1] `http://www.ncbi.nlm.nih.gov/mesh`

[2] `http://umlsinfo.nlm.nih.gov`

entities are marked explicitly. In the medical domain, Rinaldi et al (2006) and Fundel et al (2007) use manually defined relation extraction patterns based on dependency trees, and Katrenko and Adriaans (2007) apply machine learning for learning medical extraction patterns. Section 4 applies a lightly supervised approach for learning relations between named entities. Fully parsed, but otherwise unannotated, data is used. Section 5 uses a corpus annotated with medical concepts and relations to learn dependency patterns between medical concepts.

One of the main reasons for adopting dependency patterns is that it allows one to ignore intervening constituents and variations in word order that are not essential for the extraction pattern. For a language such as Dutch, this may be particularly important, as Dutch has a high degree of word order variation. The examples in (1), for instance, all contain a causal relation expressed by the phrase *wordt veroorzaakt door* (*is caused by*). If one has access to surface word order only, identification of a single extraction pattern from such data is practically impossible. Using dependency paths, on the other hand, one can extract a path linking the subject of the passive auxiliary *worden* to the object of the prepositional *door* modifier of the participle *veroorzaakt* from all these sentences.

(1) a. AIDS wordt veroorzaakt door het retrovirus HIV (AIDS *is caused by the retrovirus* HIV).
b. Nachtblindheid wordt meestal veroorzaakt door een tekort aan vitamine A (*Night blindness is usually caused by a lack of vitamin A*)
c. Echte griep of influenza is een ziekte die veroorzaakt wordt door het influenzavirus (*Real flu or influenza is a disease that is caused by the influenza virus*)
d. Buiktyfus is een geheel andere (darm) ziekte , die door Salmonella bacteriën wordt veroorzaakt (*Typhoid is a whole other (intestine) disease, which is caused by Salmonella bacteria*)
e. Brucellose bij mensen wordt met name door brucella melitensis veroorzaakt (*Brucellosis in humans is often caused by Brucella melitensis*)

The use of dependency patterns obtained from large amounts of automatically parsed data has recently also been explored for various lexical acquisition tasks, such as paraphrase learning or acquisition of taxonomic information. Lin and Pantel (2001), for instance, use 1 Gb of text parsed with Minipar (Lin, 2003), from which they extract 7M dependency paths and 200K unique paths, for learning paraphrases. Snow et al (2005) use a newswire corpus of 7M sentences, from which they extract 700K of unique noun pairs, for learning hypernyms. McCarthy et al (2007) use the written portion of the British National Corpus (90M words), parsed with RASP Briscoe and Carroll (2002), to construct a thesaurus for learning predominant word senses. Padó and Lapata (2007), finally, use all of the 100M words from the BNC parsed with Minipar for a range of lexical semantic acquisition tasks. In the experiments below, an automatically parsed version of a Dutch newspaper corpus (80M words) and a medical corpus consisting of web pages, reference works, and Wikipedia (almost 3M words) is used. More recent experiments, discussed briefly in Section 4, use a 700M word corpus.

3 Dependency Information for Question Answering and Relation Extraction

Alpino (Bouma et al, 2001; van Noord, 2006) is a wide-coverage, robust, parser for Dutch. Its grammar is designed following the ideas of a Head-driven Phrase Structure Grammar (Pollard and Sag, 1994), it uses a maximum-entropy model for statistical disambiguation, and coverage has been increased over the years by means of the semi-automatic extension of the lexicon based on error-mining (van Noord, 2004). Efficiency is improved by using a Part-Of-Speech tagger to filter out unlikely POS tags before parsing (Prins and van Noord, 2001), and by means of a technique which filters unlikely derivations based on statistics collected from automatically parsed corpora (van Noord, 2009).

Alpino is a crucial component of *Joost*, an open-domain question-answering system for Dutch (Bouma et al, 2005). Within the IMIX project, Joost was used as a QA module of the interactive, multimodal, medical QA system. Joost was also used to participate in the CLEF QA evaluation tasks, and achieved the best results for Dutch (Bouma et al, 2006).

Whereas most QA systems only use parsing to analyse the question and sometimes to analyse text snippets returned by the IR component, Alpino was used to parse the complete text collections used in the various QA systems (ranging from 2M to 110M words). The benefits are that syntactic information can be used to optimise the IR process (Tiedemann, 2005), and that off-line answer extraction can be based on dependency patterns. Jijkoun et al (2004) show, for instance, that both recall and precision of patterns for extracting answers off-line improve if patterns are dependency paths, instead of surface strings.

A successful component of many QA systems is the ability to answer questions not only by means of a method that extracts potential answers from passages returned by an information retrieval component, but also to answer questions using data that was collected by means of information extraction. For instance, if users ask frequently for the birth date of famous people, one may use information extraction to locate all instances of the *person-birth date* relation in the corpus beforehand. It has been shown that using the results of either manually constructed (Soubbotin and Soubbotin, 2002; Bouma et al, 2006), or automatically created (Ravichandran and Hovy, 2002; Fleischman et al, 2003) tables with relation instances improves the accuracy of a QA system considerably. The QA system incorporates the possibility to answer questions by means of table look-up, where the tables contain facts extracted by means of manually or automatically constructed extraction patterns. As parsed corpora are used for the QA system, extraction patterns are formulated in terms of grammatical dependency patterns.

For the information extraction experiments described below, dependency trees are used, as produced by Alpino, to extract dependency paths connecting two entities. In particular, given a pair of entities occurring in a single sentence, the dependency pattern connecting the two entities is extracted from the dependency tree for that sentence. In the implementation, a dependency pattern is the shortest path through the tree connecting two nodes, where the nodes themselves are replaced by placeholders ARG1 and ARG2.

For example, for the sentences in (2), Alpino produces the dependency trees given in Figure 1.

(2) a. Begin volgend jaar treedt ook het Spaanse Telefónica tot Unisource toe (*Early next year, the Spanish Telefónica will also join Unisource*)
b. Een gebrek aan insuline leidt tot suikerziekte (*A shortage of insulin leads to diabetes*)

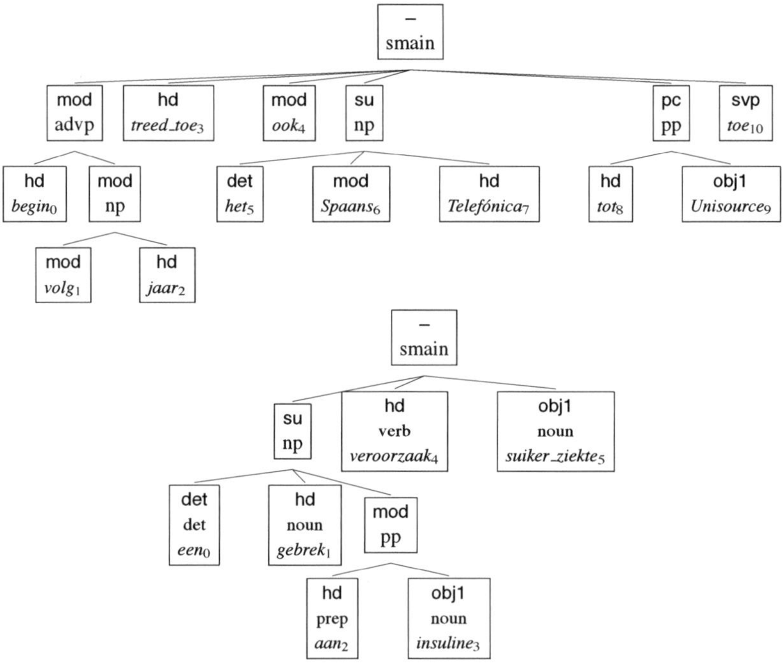

Fig. 1 Dependency tree for (2-a) and (2-b).

The dependency patterns connecting *Telefónica* and *Unisource* and *insuline* and *suikerziekte*, respectively, are:

(3) a. ARG1+*su* ← *treed_toe* → *pc+tot+obj1*+ARG2
b. ARG1+*obj1+aan+mod+gebrek* ← *leid* → *pc+tot+obj1*+ARG2

Given a dependency tree, the shortest path connecting two nodes is constructed by starting from one of the arguments, and going up in the tree until a constituent is reached that dominates the other node as well. Each time one has to go one node up, a string *rel+hd-root* is suffixed to the path expression, where *rel* is the dependency relation of the current node, and *hd-root* is the root form of the head sister node (labelled with *hd*). The same is done for the second argument, but now the expression *hd-root+rel* is prefixed to the path expression. The head root of the minimal constituent that dominates both arguments is used as a pivot expression. If one of the arguments is itself the head of a constituent dominating the other argument, this argument is itself the pivot, and one half of the pattern remains empty.

As pointed out above, one advantage of using dependency paths over patterns based on the surface string, is that dependency paths are able to deal with word order variation. Note that this is especially relevant for languages like Dutch or German, where there is considerable word order freedom, as illustrated by the (somewhat abbreviated) grammatical variants of (2-a) in (4).

(4) a. Ook Telefónica treedt begin volgend jaar tot Unisource toe
b. Ook Telefónica treedt begin volgend jaar toe tot Unisource
c. Telefónica treedt begin volgend jaar ook toe tot Unisource
d. Begin volgend jaar treedt Telefónica toe tot Unisource

For surface-based approaches, each word order variant may lead to a separate pattern, whereas this method extracts the same dependency path in each case. Another advantage is that dependency paths often capture more of the relevant context than surface patterns. Note, for instance, that the verb stem in (2-a) (*treedt*) precedes the subject, while a verbal particle (*toe*) follows the object. Surface based pattern extraction methods tend to concentrate on the string between the two arguments in a relation, and do not always capture enough of the preceding or following context to obtain an accurate pattern. Finally, note that the preceding context contains an adverb, *ook*, and the name *Telefónica* is prefixed with a determiner and a modifier (*het Spaanse*), which most likely is not relevant for formulating an accurate pattern, and thus would have to be ignored somehow.

Stevenson and Greenwood (2009) compare various methods for using dependency tree information in pattern creation for IE. Methods extracting only *subject-verb-object* tuples have limited coverage, whereas methods extracting the minimal subtree containing both arguments suffer from lack of generality. Their *linked chain* method corresponds to the shortest path pattern extraction method, and performs well in an evaluation using the Wall Street Journal and biomedical data.

4 Relation Extraction for Open Domain QA

This section presents a simple lightly supervised information extraction algorithm which operates on a parsed corpus to learn dependency patterns for extracting relation instances. The primary goal is to use this system for off-line extraction of instance pairs that can be used to provide answers for frequently asked question types. Thus, evaluating the results of the extraction process not only in terms of precision, but also by integrating them into the QA system *Joost*.

The algorithm takes as input seed pairs representing a particular question category. For example, for the category of *capital-of* questions it is fed ten *country-*

capital pairs like *Germany-Berlin*. From each sentence in which a seed pair occurs, the dependency path connecting the two elements is extracted. Extraction patterns are selected by estimating the precision of each dependency path, and preserving only those paths that are above a given threshold.

The results of the experiments indicate that, for QA, an extraction method that aims for high recall (possibly at the expense of precision) gives the best results. The performance of the QA system, although sometimes tables of relation instances are extremely noisy, improves most if high coverage is used.

4.1 Pattern Induction

The starting point is a number of seed pairs. Extraction patterns are found by searching the corpus exhaustively for all sentences in which both arguments of a seed pair occur. The shortest dependency path between the two arguments is selected as an extraction pattern. For instance, given a seed pair *Letland, Riga*, the pattern (5-b) for the sentence (5-a) is constructed.

(5) a. Riga, de hoofdstad van Letland, is een mooie stad (*Riga, the capital of Latvia, is a beautiful city*)
    b. ARG2 → *app+hoofdstad+mod+van*+ARG1

A given seed set typically gives rise to a large set of dependency patterns, especially if some of the seed pairs are frequent in the corpus. However, not all dependency patterns are adequate as an extraction pattern. For instance, sentence (6-a) gives rise to the highly general pattern (6-b). Including this pattern as an extraction pattern would mean that many false instance pairs are extracted from the corpus in the extraction stage.

(6) a. De in Riga in Letland geboren Leibowitz studeerde in Berlijn (*Leibowitz, born in Riga in Letland, studied in Berlin*)
    b. ARG1 → *mod+in*+ARG2

In the pattern-filtering stage, unreliable patterns are filtered. To find the optimal balance between precision and recall for off-line answer extraction three different bootstrapping experiments are performed in which the precision threshold for selecting patterns is varied. The precision of a pattern is calculated following the method in Ravichandran and Hovy (2002). Instead of replacing both seed terms by a variable as in (5-b), only the answer term (i.e. of *what is the capital of... ? questions* is replaced with a variable:

(7) ANSWER → *app+hoofdstad+mod+van+Letland*

For each answer pattern obtained in this way, how many times the pattern occurs in the corpus is counted, and how many times the variable ANSWER matches with the correct answer term according to the seed list is counted. The precision of each pattern is calculated by the formula

$$P = C_a / C_o$$

where C_o is the total number of times the pattern occurred in the corpus, and C_a is the total number of times the pattern matched and the ANSWER matched the correct answer term.

All patterns that occur at least two times in the corpus, and that have a precision score exceeding a set threshold τ_p, are preserved for the instance extraction phase. The patterns that have passed the filter in the previous stage are matched against the parsed corpus, to retrieve new relation instance pairs. After retrieval, a random sample is selected of one hundred instance pairs and manually evaluated. If more than τ_f facts are found, the iteration process is stopped. Alternatively, all facts are used again, without any filtering, as seeds for the pattern-induction stage and the process repeats itself. In the experiments, τ_f is set to 5000.

4.2 Experiment

Experiments were performed using the CLEF corpus. This is a 80M word consisting of newspaper articles from 1994 and 1995. It is the corpus that is used for Dutch in the QA task of CLEF. Two question types are selected that are frequent in the CLEF QA question set[3] (Magnini et al, 2003): *capital-of* a *soccer-player-club*. These are binary relations, respectively between a location (e.g. *France*) and its capital (*Paris*) and between a soccer player (e.g. *Dennis Bergkamp*) and his club (*Ajax*). [4]

The *capital-of* relation if functional, i.e. for a given country there is only a single capital. The *soccer-player-club* relation is not one-to-one, as a club has many players, and players can also be playing for more than one club during the two year period covered by the CLEF corpus.

For each of the two question types ten seed facts are created which are listed in Table 1. The initial seeds were chosen with some care since they form the basis of the learning process. For instance, not only national capitals were included, but also the capital of a Dutch province (*Drenthe*) and *Brussels*, the capital of *Europe*. Both the adjectival form of a country name as well as the name itself to cover a greater variety of patterns were included. For the football seeds it was ensured that the seeds were instances that were true for the period 1994-1995, and both full names and last names were included.

Only patterns with a frequency higher than one were selected. In the most lenient experiment all these patterns were used to extract new facts in the CLEF corpus. Three experiments were performed: retaining all patterns (i.e. $\tau_p = 0.0$); retaining only patterns with a precision $P \geq 0.5$ ($\tau_p = 0.5$); and retaining only patterns with

[3] `http://clef-qa.itc.it/2005/resources.html`

[4] Mur (2008) also presents results for learning the ternary *minister-country-department* relation.

Table 1 Ten *capital-of* and *soccer-player-club* seeds.

Country	Capital	Person	Club
Amerikaanse	Washington	Litmanen	Ajax
Bulgaarse	Sofia	Marc Overmars	Ajax
Drenthe	Assen	Wim Jonk	Inter
Duitsland	Berlijn	Dennis Irwin	Manchester United
Europese	Brussel	Desailly	AC Milan
Frans	Parijs	Romario	Barcelona
Italiaans	Rome	Erwin Koeman	PSV
Bosnisch	Sarajevo	Jean-Pierre Papin	Bayern München
Rusland	Moskou	Roberto Baggio	Juventus
Spaans	Madrid	Aron Winter	Lazio Roma

Table 2 Sample of question used for evaluation.

Question	Answers
Wat is de hoofdstad van Canada?	Ottawa
Wat is de hoofdstad van Cyprus?	Nicosia
Wat is de hoofdstad van Haïti?	Port-au-Prince
Bij welke club speelt Andreas Brehme?	1. FC Kaiserslautern
Bij welke club speelt Aron Winter?	Lazio Roma; Lazio
Bij welke club speelt Baggio?	Juventus; Milan

$P \geq 0.75$, where P is computed as described above. This process is repeated for two iterations or until more than 5,000 relation instance pairs are found. Note that each matching occurrence of an instance pair in the corpus is individually counted. The reason for this is that for QA it is important to be able to *justify* an answer: i.e. for a given question, not only the answer should be provided, but also the sentence or paragraph from which it was extracted. An answer is *justified* only if this surrounding context provides support for the truth of the answer.

After the extraction stage, tables are obtained that can be used to provide answers for *Capital-of* and *soccer player* questions. To test the effect of including these tables in the QA system *Joost*, the number of relevant questions was expanded in the CLEF QA test sets with a number of questions that were self created. *Capital-of* questions were googled for *Wat is de hoofdstad van (What is the capital of)* to find more of them. Football questions were created by asking five people names of famous football players in 1994 and 1995. For each name in the responses, a question of the form *Bij welke club speelde X? (For which club did X play?)* was created. All questions were checked to ensure that an answer was present in the CLEF corpus. In the end detail tests were carried out on 42 capital questions and 66 football questions. A few example questions with their answers are given in Table2.

4.3 Evaluation

The extraction relation instances were evaluated by estimating their precision, and by incorporating them in *Joost* as tables that can be used to provide answers to user questions. The precision was estimated based on a random sample of 100 relation instance pairs. To evaluate the results of the QA system, the number of times the first answer was correct was simply counted and the mean reciprocal rank over the first 5 answers was computed. The reciprocal rank score for a question is $1/R$, where R is the rank of the first correct answer.

The results are given in Tables 3 and 4. Using no filtering (i.e. $P \geq 0.0$), 39 patterns for *capital-of* in the first round (i.e. using only the seeds given in Table 1) were found. Applying these 39 patterns 3,405 new instance pairs were extracted. The estimated precision is 0.58. When using these instance pairs as a table for off-line question answering in *Joost*, 35 of 42 questions were answered correctly. The mean reciprocal rank was 0.87. For the second round the process was repeated using 3,405 instance pairs. With these facts 1,306 patterns were found. The 1,306 patterns in turn returned 234,684 instance pairs.

The middle and bottom part of the tables show the results for the experiment in which the patterns are filtered using $P \geq 0.5$ and $P \geq 0.75$. For the *capital-of* relation it was stopped after two iterations and for the *soccer-player-club* relation it was stopped after it found more than 5,000 facts. The best performance per category is marked in bold.

Table 3 Results for learning patterns for answering *capital-of* questions.

Capital-of $P \geq 0.0$	# patterns	# pairs	(P)	1st ans OK (# q = 42)	MRR
1st round	39	3,405	(0.58)	35	0.87
2nd round	1306	234,684	(0.01)	14	0.51
$P \geq 0.5$					
1st round	24	2,875	(0.63)	35	0.85
2nd round	171	4,123	(0.49)	**37**	**0.90**
$P \geq 0.75$					
1st round	17	2,076	(0.83)	35	0.84
2nd round	64	2,344	(0.83)	35	0.85

The results for using automatically created relation instance tables in the QA system show that even low precision data can help to improve the performance of the QA system. The best results for answering *capital-of* questions were obtained in the second round with $P \geq 0.50$. The precision of the extracted facts was only a mediocre 0.49 compared to 0.83 in the experiments with $P \geq 0.75$. The number

Table 4 Results for learning football patterns (Best results in bold).

Football $P \geq 0.0$	# patterns	# pairs	(*P*)	1st ans OK (# q = 66)	MRR
1st round	19	115,296	(0.01)	40	0.66
$P \geq 0.5$					
1st round	11	109,435	(0.01)	**41**	**0.67**
$P \geq 0.75$					
1st round	6	196	(0.26)	11	0.17
2nd round	28	31,958	(0.02)	18	0.31

of facts, on the other hand, was almost twice as high (4,123 vs. 2,344). The QA evaluation shows that 'recall' is more important than precision in this case.

Table 4 illustrates this effect even more strongly. The best result is again $P \geq 0.50$, but this time there is no significant difference with the result computed for the experiment where no pattern filtering was applied. An extremely large number of incorrect instance pairs were extracted in both cases, but this did not hurt performance on the QA task.

This rather contradictory result can be explained by the patterns that were found for the extraction of football facts. A very frequent pattern found was *Player* $\rightarrow$ *mod+Club*. The pattern occurs typically when a name is followed by a name in brackets, i.e. it matches for example with the phrase *Jari Litmanen (Ajax)* but also with *Rijksuniversiteit Groningen (RUG)*. Although the pattern is noisy, the incorrect facts typically have nothing to do with football players, and thus they do not cause incorrect answers to football questions (where the name of the player is always given).

The results of the experiments for the extraction of *capital-of* and *soccer player-club* instance pairs suggest that for the benefit of off-line QA it is better to focus on high recall than on high precision. The use of a pattern filtering method based on the estimated precision of patterns provides a method for balancing the precision and recall of the extraction process. It should be noted, however, that this is a rather crude method. The experiments illustrate that varying the value of *P* used for filtering can easily lead to a situation where either very few instance pairs are extracted or where an excessive number of instances are extracted, with very low precision. The latter situation makes further iterations of the extraction process fruitless (as the level of noise is simply too high). A second problem for the method above is the fact that all relation instances found in iteration I are used as seeds for iteration $I+1$. Given the low precision of some of the experiments, it becomes interesting to search for methods that use as seeds only the most reliable instances from a previous round.

The *Espresso* algorithm of Pantel and Pennacchiotti (2006) is a lightly supervised information extraction method that offers a better balance between precision and recall. By selecting only the most reliable patterns and only the most reliable relation instances in each round of the bootstrapping process, more iterations can be carried out without large drops in precision. The reliability of instances and patterns is computed by means of a scoring criterion based on the mutual information score between instance pairs and patterns. Recent work by Ittoo and Bouma (2010) and Bouma and Nerbonne (2010) uses *Espresso* for relation extraction on a 700M word corpus, containing among other things the CLEF corpus. Whereas Pantel and Pennacchiotti (2006) use surface strings as extraction patterns, the experiments described here use dependency patterns for extraction.

Using the seed list for the *soccer-player-club* relation given in Table 1, and using the *Espresso* algorithm for relation extraction, the results were obtained in Table 5. The first two columns give results for using the method of Pantel and Pennacchiotti (2006). As in that paper, initially the 10 highest scoring patterns are selected, and 100 instance pairs and 1 pattern per iteration are added. Precision in all iterations is as good or better as that of the experiment with the highest precision in Table 4. As with the experiments above, however, iterative, lightly supervised methods like this are subject to *semantic drift*, i.e. the phenomenon that errors in previous iterations have a deteriorating effect on the accuracy of later iterations (McIntosh and Curran, 2009). To dampen this effect, distributional similarity (Lin, 1998; van der Plas, 2008) was used to filter instance pairs where the first element is not distributionally similar to the group of *soccer players* or where the second element in not similar to *soccer clubs*. The results for this method are given in the final two columns.

Table 5 Accuracy per iteration for learning the *soccer* relation using *Espresso* and *Espresso* combined with a distributional similarity filter.

	Espresso		Espresso^{+}	
	pairs	prec	pairs	prec
1st round	109	0.36	40	0.65
2nd round	211	0.30	74	0.53
3rd round	312	0.25	88	0.38
4th round	412	0.27	176	0.41
5th round	511	0.33	290	0.45

It is hard to compare the recall of the technique used for creating tables for QA and the relation extraction method based on *Espresso*. Most importantly for QA, individual occurrences of instance pairs are counted (the context of each instance pair is needed as *justification* of the answer). For general purpose relation extraction, on the other hand, all occurrences of an instance pair are counted as a single instance. If individual occurrences are counted, the number of instances retrieved for the *Espresso* experiments is almost 12,000, whereas for the system with filtering it is over 7,000. This suggests that recall might still be sufficient for integration of the *Espresso* technique in a QA system, but this had not been tested at the time of writing.

5 Relation Extraction for Medical QA

Relation extraction for open domain QA has concentrated on relation types where the arguments are named entities and dates. In the large corpora used for open domain QA, these occur relatively frequently, with little variation in spelling. Thus, lightly supervised methods, that rely on the fact that many snapshots of a relation are present in the corpus, and that some of these are highly frequent, has been used successfully as a component in open domain QA. For closed-domain, medical, QA, the situation is more complex. Here, the relations of interest typically exist between concepts expressed as complex noun phrases. An example is given in (8).

(8) De ziekte van Graves-Basedow (toxische diffuse struma) wordt vermoedelijk veroorzaakt door een antilichaam dat de schildklier aanzet tot overproductie van het schildklierhormoon. (*Graves' disease (toxic diffuse goitre) is most likely caused by an anti body which leads the thyroid to excessive production of the thyroid hormone.*)

This sentence expresses a causal relationship between a disease (*De ziekte van Graves-Basedow* and a cause, *een antilichaam dat de schildklier aanzet tot overproductie van het schildklierhormoon.* It is unlikely that these exact two phrases will ever occur frequently as a pair in the corpus. As this is true for most of the instances of the *cause* relation in the corpus, the chances of bootstrapping extraction patterns from seeds for the cause relation are not very promising. Most work on medical relation extraction has therefore used at least some amount of annotated data. This describes how relation instances can be extracted from a Dutch medical corpus, using annotated data and UMLS to guide the extraction process.

Within the IMIX project, a substantial corpus of Dutch medical text has been annotated with semantic labels.[5] The annotated corpus (approx. 600K words) consists of texts from a medical encyclopedia and a medical handbook. An example of the annotation is given in Figure 2.

```
<rel_causes>
   Een tekort aan
          <con_body_part>insuline</con_body_part>
   leidt tot
          <con_disease>suikerziekte</con_disease>
</rel_causes>
```

Fig. 2 Semantic annotation in the IMIX corpus (of the sentence *A shortage of insuline leads to diabetes*).

[5] The corpus, developed by the University of Tilburg IMIX/Rolaquad project (`ilk.uvt.nl/rolaquad`) is also used in the chapter by Canisius, van den Bosch, and Daelemans, *Constraint-Satisfaction Inference for Entity Recognition* and in the chapter by Lendvai, *Towards a Discourse-driven Taxonomic Inference Model.*

Sentences may be labelled with relation tags, such as *rel_causes*. Noun phrases denoting concepts are annotated with one of 12 semantic concept types such as *body_part* or *disease*. Seven different relation types are present in the corpus: *causes, has_definition, diagnoses, occurs, prevents, has_symptom* and *treats*). The frequency of each relation ranges from 479 (*prevents*) to 4,301 (*treats*). Some sentences are annotated with more than one relation type. This annotation is very helpful for learning medical relation extraction patterns, although it should be noted that no labelling is present which explicitly identifies the arguments of a relation. Furthermore, it is not guaranteed that suitable arguments for a relation can actually be found within the sentence. For instance, in (9), the subject *dit alles* (*all this*) is an anaphoric expression which is not in itself a suitable argument for a medical relation instance pair.

(9) Dit alles duidt op een verschil in ontwikkeling van de hersenen bij jongetjes en meisjes (*All this indicates a difference in the development of the brain in boys and girls*)

The corpus was used to learn dependency patterns that are associated with a given medical relation. It turns out to be the case that many relations can be expressed in text by general linguistic patterns (*X may lead to Y, X occurs in Y*), which are not unique to a given medical relation, and also, which do not imply that both *X* and *Y* are medical concepts. Such patterns can nevertheless be used to extract medical relations with high accuracy if it is required that both *X* and *Y* are medical terms. The restriction may also be imposed that *X* and *Y* have to be terms that belong to a given class (i.e. *X* and *Y* are medical terms denoting, respectively, a *virus* and a *disease*). By imposing restrictions on the semantic class of the argument, the ambiguity of the dependency patterns were also reduced.

Below, a method for predicting concept labels of Dutch medical terms, using (English) UMLS as a thesaurus was first presented. Next, the method for learning extraction patterns based on the IMIX corpus was presented. Then it was shown that the combination of extraction patterns and semantic concept labels provided accurate results for relation extraction. Finally, the effect of incorporating the extracted relation instances in a medical QA system was evaluated.

5.1 Multilingual Term Labelling

Medical terminology differs across languages, but it is also closely related. Technical medical terms in Dutch, for instance, often are simply borrowed from English (i.e. *stress, borderliner, drugs* and acronyms like ADHD and PTSS), or are cognates (i.e. English *genetic* and Dutch *genetisch*). Some terms are genuinely different in the two languages (*infection* and *besmetting*), and need to be translated.

For classifying Dutch terms on the basis of a subset of UMLS concepts that contain Dutch and English terms,[6] a sequence of five heuristics is used, illustrated in Table 6. If step 1 returns no result, step 2 will be evaluated, and so on. For the example *psychische aandoeningen (mental disorders)*, step 3 returns the result *B2.2.1.2.1.1:Mental or Behavioral Dysfunction*. In case there is more than one result (this happens especially in steps 4 and 5), a further heuristic is needed to decide the best labels. Given a query term QT, a match may be found with a term in UMLS UT, which has concept type UC. If a query returns more than three results, the results are restricted to cases where QT and UT have the same length. If more than three results remain, all Dutch terms UT are filtered which have the head in a different position than QT. Finally, the remaining semantic types UC are ranked by frequency, and the three highest ranked types are selected.

Table 6 Five conditional steps in classifying a Dutch term, *psychische aandoeningen* (*'mental disorders'*), based on a subset of UMLS concepts in Dutch (NL) and English (EN).

No	Lang	Index	Query parameters	Example
1	NL	Root	Exact match of root forms	Psychische aandoening
2	NL	Term	Exact match of term string	Psychische aandoeningen
3	EN	Term	Exact match of translated term[7]	Mental disorders
4	NL	Head	Exact match of head word	Aandoening
5	NL or EN	Term	One of the words in the term	Psychisch OR mental OR ...

The contribution of each of the heuristics was evaluated by applying the method pages in the category Health Care of Dutch Wikipedia. 370,578 terms were found. 17% of these were found in the Dutch part of the UMLS and 30% in the English part (through translation); 38% are new terms which could be assigned a concept label in steps 4 and 5, and 16% of the terms received no label. For the new terms, 26% were labelled using heuristic 4 (matching the Dutch head word) and 74% using heuristic 5 (a matching Dutch or English word). This shows that labelling new Dutch terms benefits from reusing labels of existing terms, and from translation.

The accuracy of the method was evaluated on the 1000 most frequent terms in the IMIX medical corpus. As the system returns (one of 135) UMLS semantic concepts, whereas the IMIX corpus uses (only 12) corpus-specific concept labels, a (many to one) mapping from UMLS labels to more corpus concept labels was defined. Note that each UMLS label was mapped to, at most, one corpus label. Evaluation is carried out on the basis of the highest ranked UMLS concept assigned by the heuristics outlined above. A precision of 78.2% was obtained.

It should be noted that the mapping creates certain mismatches. For instance, the term *tandsteen* (*calculus*) was labelled as *disease* in the corpus but as *Body Substance* in UMLS (and by the classification method). It is believed both classes

[6] The relevant subset of UMLS consists of 163,032 concepts in Dutch and 2,974,889 concepts in English.

are correct, although in general it is not correct to map *Body Substance* to *disease*. Thus, one suspects that the actual accuracy of the method may be slightly higher than the precision figure suggests.

Canisius et al (2006) use a machine learning approach to train a concept classifier for the same data. They do not use external resources, but instead try to learn the classification from (a subset of) the corpus itself. They report an accuracy of 68.9% and a theoretical upper bound of 74.9%. This suggests that, given the limited size of the corpus, the use of external knowledge sources (UMLS in particular) boosts the performance of concept labelling.

5.2 Learning Patterns

Given two medical terms in a sentence labelled with a medical relation, the shortest dependency path is extracted connecting the two as an extraction pattern. For instance, given sentence (10-a), the patterns (10-b) and (10-c) may be extracted amongst others .

(10) a. Aantasting van de bijnierschors door infecties (bijv. tuberculose) of bij auto-immuunziekten kan leiden tot de ziekte van Addison (*Erosion of the adrenal glands by infections (e.g., tuberculosis) or with autoimmune diseases could lead to Addison's disease*)
b. ARG1+*obj1+van+mod+aantasting+subj* ← *leid* → *pc+tot*+ARG2
c. ARG1+*subj* ← *leid* → *pc+tot*+ARG2

Both ARG1 and ARG2 are required to match a medical term. In (10-a) the case for *aantasting van de bijnierschors* and *ziekte van Addison*, and thus one of the patterns obtained was:

(11) NEOPLASTICPROCESS+*subj* ← *leid* → *pc+tot*+DISEASEORSYNDROME

(11) is an example of a *semantic extraction pattern*, i.e. a dependency pattern with semantic (UMLS) class labels for ARG1 and ARG2. To find the appropriate semantic label for a complex argument, first its main term is extracted using a linguistic filter adapted from (Justeson and Katz, 1995). The filter extracts a substring of the argument that matches the following POS-tag regular expression:

(12) `((Adj | N)* N Prep Det?)? (Adj | N)* N`

For the subject in the example above, it extracts *Aantasting van de bijnierschors* (`N Prep Det N`) as the main term. Next, the semantic class labels are found for the term using the method outlined in Section 5.1.

The task of pattern learning is to find sets of semantic extraction patterns for each of the relations in the IMIX corpus. For each pair of medical concept terms in a sentence, the dependency path connecting the two is extracted. For the main terms in the concepts, the UMLS class labels are determined. Where this returns more than one concept label, all combinations of concept labels are used to generate semantic extraction patterns. For instance, for (10-a) the following semantic relation patterns are obtained:

(13) a. DISEASEORSYNDROME+*subj* ← *leid* → *pc*+*tot*+DISEASEORSYNDROME
b. NEOPLASTICPROCESS+*subj* ← *leid* → *pc*+*tot*+DISEASEORSYNDROME
c. FINDING+*subj* ← *leid* → *pc*+*tot*+DISEASEORSYNDROME

Dependency patterns are ranked according to the relative frequency with which they occur with a given relation and patterns below a certain threshold are discarded. For dependency pattern ranking, the weight of a pattern R is computed as the ratio of the probability $P(R_C)$ with which R was found in a training corpus C containing only sentences labelled with the relevant relation, and its probability $P(R_G)$ in the general medical corpus G. This score is multiplied with the frequency of R in C, as shown below:

$$weight(R) = \frac{P(R)}{P(R_G)} \times f(R_C)$$

The intuition behind this method is that good patterns ought to appear more frequently in the training corpus for the relation than in the general corpus. Multiply with frequency again to decrease the importance of low frequency patterns in the training corpus.

Semantic relation patterns consist of dependency patterns extended with semantic labels. Semantic relation patterns $R_{(A,B)}$ are weighted by multiplying their frequency with the weight of their *dependency patterns R*:

$$weight(R_{(A,B)}) = weight(R) \times f(R_{(A,B)})$$

During relation extraction for a given relation *Rel*, all relation instances *R(*ARG1, ARG2*)* are extracted from a corpus, where the dependency pattern R has to be a valid dependency pattern for *Rel* and the semantic types of `Arg1` and `Arg2` have to match one of the semantic relation patterns $R_{(A,B)}$ for *Rel*.

To investigate the effect of concept labelling, relation extraction experiments are also carried out where both arguments of a potential relation instance did not match the semantic types of a semantic relation pattern, but instead only one or no argument matched.

5.3 Evaluation

The accuracy of the semantic relation extraction patterns are evaluated on a subset of the IMIX corpus, and on the text from the Health Care section of Wikipedia.

From the IMIX corpus, for each relation, 50 sentences as a test set were randomly selected. Note that these were withheld from the corpus that was used for learning the extraction patterns. From these 50 sentences only those were selected that contain two fully specified arguments of the relation and these were discarded from the test set sentences containing e.g. anaphoric NPs as an argument. Note that, since relation labelling was done at the level of sentences, for many relations, less than 50% of the labelled instances actually contain both arguments of the relation. This indicates that the corpus is a good deal less informative than corpora which explicitly mark relations between (medical) terms. Table 7 gives the results for the various relations.

Table 7 The relation extraction results on the test data of 50 sentences per relation type. #pat is the number of semantic extraction patterns for that relation.

Relation type	# pat.	P	R	F
Has_definition	4	0.83	0.73	0.78
Causes	155	0.92	0.67	0.77
Occurs	71	0.81	0.54	0.65
Has_symptom	206	0.58	0.62	0.60
Prevents	47	0.80	0.40	0.53
Treats	180	0.71	0.40	0.51
Diagnoses	85	0.86	0.24	0.38

Precision (i.e. the number of times the relation label *R* are correctly predicted divided by the number of times this label was predicted by the system) is relatively high for all patterns, but recall varies. The method performs reasonably well for the `has_definition` (f-measure .78) and `causes` (.77) relation types, and performs less well for the `diagnoses` relation (.38). Variation in performance is probably due to the fact that for some relations, more training examples are available, some relations are expressed by simpler dependency patterns (i.e. *is caused by*), and some patterns seem to suffer more from parsing errors.

The relations `causes` and `has_symptom` have many similar dependency patterns (*be_cause_by*). However, the semantic classes of their arguments are different. In the first ten relation patterns for `causes` the object argument has diverse semantic types: *Disease or Syndrome* (3), *Finding* (2), *Pathologic Function*, *Functional Concept*, *Protein*, *Bacterium*, and *Virus*, while `has_symptom` is dominated by *Sign or Symptom* (6). This shows how the semantic type of a term, along with its pattern, plays an important role in identifying the type of a relation instance.

To further evaluate the effect of using semantic types in the relation extraction task, the patterns on a subset of Wikipedia text were tested that contained medical lemmas. 20 extraction results for each relation and each level of matching of the semantic argument types were randomly selected, and were manually evaluated. The number of results for each level and the precision is given in Table 8.

For all of the relation types, the best accuracy is at level 2M, where both of the arguments have semantic types matching a pattern for that relation. The drop in precision at level 0M (where no matching argument was found) is considerable. For

Table 8 Number of matching dependency relation patterns per relation with 2, 1 , or 0 matching concept labels.

Relation type	2M #	2M prec	1M #	1M prec	0M #	0M prec
Causes	942	0.95	1,625	0.90	647	0.75
Has_definition	4,102	0.95	6,118	0.65	657	0.30
Occurs	548	0.90	2,219	0.80	1,238	0.50
Treats	300	0.85	1,826	0.60	1,026	0.45
Has_symptom	1,220	0.80	2,668	0.30	850	0.00
Prevents	24	0.75	171	0.50	470	0.50
Diagnoses	34	0.60	265	0.60	231	0.35
Total	7,170	0.83	14,892	0.62	5,149	0.41

the `prevents` relation type, 50% of the errors at 1M are cases where one of the arguments is a definite NP. To obtain a full interpretation of these NPs, they need to be interpreted as coreferential with a preceding NP. Although the generalisation of patterns increases recall, it also becomes the main cause of errors at 1M and, especially, 0M. The noise in the training data also contributes to errors, although the pattern filtering has reduced a great amount of irrelevant patterns. Another source of errors are non-medical term arguments, such as the names of places, concepts for other domains, or non-term arguments.

The low accuracy scores at 0M are a further indication that the coverage of the concept labelling system is satisfactory: if both of the arguments cannot be assigned a medical concept label, that the extracted arguments are not proper instances of the relation is relatively certain.

5.4 Evaluation in a QA *Setting*

In this section, the performance is reported of the QA system *Joost* on a test suite of questions from the medical domain. The test suite was derived from the pool of 435 candidate questions by Tjong Kim Sang et al (2005), and expanded with questions found on the web (by submitting keywords and phrases from typical medical questions to a search engine). Many of the candidate questions found on the web have no answer in the IMIX medical corpus. Only questions which have at least one answer in the corpus were selected.

The performance of the QA system on 58 questions was tested, covering three question types: *has_definition* (25 questions), *causes* (22 questions) and *has_symptom* (11 questions). The results of the performance of the QA system in the three experimental settings are shown in Table 9. Here, *manual* refers to the QA system using tables created using manually constructed extraction patterns, *learned* refers to the system using tables based on automatically learned extraction patterns (as described above), and IR refers to the QA system without any tables but relying solely on passage retrieval and answer extraction. Results are given in terms of mean reciprocal rank MRR (i.e. the mean of $1/R$, where R is the rank of the first correct answer) and *1st* correct.

Table 9 Performance scores of the QA system on the three experiment settings (manual, learned, and IR), measured using MRR and first answer correct, for the three question types (*has_definition*, *causes*, and *has_symptom*).

Method	Answered	MRR	1st
Has_definition			
Manual	16	0.333	0.280
Learned	**22**	**0.465**	**0.360**
IR	21	0.133	0.040
Causes			
Manual	19	0.547	0.409
Learned	**20**	**0.750**	**0.682**
IR	19	0.405	0.318
Has_symptom			
Manual	5	0.364	0.364
Learned	8	**0.636**	**0.636**
IR	8	0.182	0.182
Overall			
Manual	40	0.420	0.345
Learned	**50**	**0.605**	**0.534**
IR	48	0.246	0.172

In general, the MRR and the first answer correct scores of the manual and the pattern learning methods outperform the scores of the IR baseline. For all of the question types, the performance scores of the system using automatically learned extraction patterns outperform the scores of the manual method. And overall, the method contributes 42% and 52% improvement against the manual method with respect to the MRR score and the first correct answer respectively. This shows that the method has successfully improved the performance of the medical QA system.

6 Conclusions and Future Work

In this chapter, the importance of relation extraction for boosting the performance of QA systems has been stressed. This is true for both open-domain QA and specialised QA such as medical QA. The experiments in open-domain QA have concentrated on methods that have high recall, sometimes at the expense of precision. In the context of a QA system, much of the noise incorporated by high coverage extraction patterns never surface, as user questions always supply one argument of the relation, and also, because the frequency with which pairs are found is used to rank answers. Correct answers tend to be extracted more often than incorrect ones, even in systems that introduce substantial levels of noise.

The experiments on medical relation extraction cannot rely on the fact that (seed) instance pairs are frequent in the corpus, and that arguments for a given relation are easily found. To overcome the problem of identifying terms denoting medical terms, a method for assigning UMLS concept labels are presented to Dutch medical terms which employ both the Dutch and the English part of the UMLS. Both the coverage and the precision of the term classification method was shown to be relatively high compared with other methods that do not use external knowledge. This experiment also shows that it is possible to use a multilingual resource to classify new terms in a particular language.

Relation patterns for medical relation extraction can be obtained from sentences that were labelled with only the relation they contain. Concept labelling helps to improve the accuracy of relation extraction: it is used to rank relevant patterns higher, to distinguish identical dependency patterns for different relations, and to predict which matching patterns in a test corpus are the most likely correct instances of the relation.

The current method uses the semantic types and semantic relation (patterns) from the training data. In the future it is planned to (semi-automatically) annotate corpora using the UMLS Semantic Network that contains 135 semantic types and 54 relationships.

The relative success of using dependency patterns with concept labels in the medical domain suggests that similar methods might also provide a means to improve the accuracy of open-domain relation extraction. In particular, term identification might help to detect term variation, also for person, organisation, and geographical names, and could help to find multiword terms. Concept labelling could help to reduce the level of noise in the current open-domain relation extraction system.

The evaluation of the medical relation extraction results noticed an important source of errors due to coreference. Sentences such as *This form is transferred via a dominant gene*, or *The disease is caused by a surplus of growth hormone* are labelled as *cause*, but were discarded as *cause* sentences from the gold standard evaluation set, as they do not contain complete information for one of the arguments of the relation. It was estimated that approximately 9% of the relation candidates in the Wikipedia data contains a pronominal or definite NP that needs anaphoric interpretation. An obvious next step would be to apply coreference resolution to medical terms, so as to obtain a full interpretation of the term, and a term which can be used for concept classification.

References

Bodenreider O (2004) The unified medical language system (UMLS): integrating biomedical terminology. Nucleic Acids Research 32(Database Issue):D267

Bouma G, Nerbonne J (2010) Applying the espresso-algorithm to large parsed corpora. Submitted.

Bouma G, van Noord G, Malouf R (2001) Alpino: Wide-coverage computational analysis of Dutch. In: Computational Linguistics in The Netherlands 2000, Rodopi, Amsterdam

Bouma G, Fahmi I, Mur J, van Noord G, van der Plas L, Tiedeman J (2005) Linguistic knowledge and question answering. Traitement Automatique des Langues 2(46):15–39

Bouma G, Mur J, van Noord G, van der Plas L, Tiedemann J (2006) Question answering for dutch using dependency relations. In: Peters C (ed) Accessing Multilingual Information Repositories, pp 370–379, URL http://dx.doi.org/10.1007/11878773_42

Braun L, Wiesman F, van den Herik J (2005) Towards automatic formulation of a physician's information needs. In: Proceedings of the Dutch-Belgian Information Retrieval Workshop, Utrecht, the Netherlands

Briscoe T, Carroll J (2002) Robust accurate statistical annotation of general text. In: Proceedings of the 3rd International Conference on Language Resources and Evaluation, Citeseer, pp 1499–1504

Bunescu R, Mooney R (2005) A shortest path dependency kernel for relation extraction. In: Proceedings of the conference on Human Language Technology and Empirical Methods in Natural Language Processing, Vancouver, pp 724–731

Canisius S, van den Bosch A, Daelemans W (2006) Constraint satisfaction inference: Non-probabilistic global inference for sequence labelling. In: Proceedings of the EACL 2006 Workshop on Learning Structured Information in Natural Language Applications, Trento

Culotta A, Sorensen J (2004) Dependency tree kernels for relation extraction. In: 42nd Annual Meeting of the Association for Computational Linguistics (ACL), Barcelona, Spain

Etzioni O, Cafarella M, Downey D, Popescu A, Shaked T, Soderland S, Weld D, Yates A (2005) Unsupervised named-entity extraction from the web: An experimental study. Artificial Intelligence 165(1):91–134

Fleischman M, Hovy E, Echihabi A (2003) Offline strategies for online question answering: Answering questions before they are asked. In: Proc. 41st Annual Meeting of the Association for Computational Linguistics, Sapporo, Japan, pp 1–7

Fundel K, Küffner R, Zimmer R (2007) Relex - relation extraction using dependency trees. Bioinformatics 23:365–371

Ittoo A, Bouma G (2010) Mereological and meronymic relations for learning part whole relations. In: Computational Linguistics in the Netherlands 2010, Utrecht, the Netherlands

Jijkoun V, Mur J, de Rijke M (2004) Information extraction for question answering: Improving recall through syntactic patterns. In: Coling 2004, Geneva, pp 1284–1290

Justeson J, Katz S (1995) Technical terminology: some linguistic properties and an algorithm for identification in text. Natural language engineering 1(01):9–27

Katrenko S, Adriaans P (2007) Learning relations from biomedical corpora using dependency trees. In: Tuyls K, Westra R, Saeys Y, Nowé A (eds) Knowledge Discovery and Emergent Complexity in BioInformatics, Lecture Notes in Bioinformatics. LNBI, vol. 4366, Springer

Lin D (1998) Automatic retrieval and clustering of similar words. In: Proceedings of COLING/ACL, Montreal, pp 768–774

Lin D (2003) Dependency-based evaluation of MINIPAR. In: A Abeillé, Treebanks: Building and Using Parsed Corpora, Kluwer, pp 317-329

Lin D, Pantel P (2001) Discovery of inference rules for question answering. Natural Language Engineering 7:343–360

Lita L, Carbonell J (2004) Unsupervised question answering data acquisition from local corpora. In: Proceedings of the thirteenth ACM international conference on Information and knowledge management, ACM, p 614

Magnini B, Romagnoli S, Vallin A, Herrera J, Peñas A, Peinado V, Verdejo F, de Rijke M (2003) The multiple language question answering track at clef 2003. In: Peters C (ed) Working Notes for the CLEF 2003 Workshop, Trondheim, Norway

McCarthy D, Koeling R, Weeds J, Carroll J (2007) Unsupervised acquisition of predominant word senses. Computational Linguistics 33(4):553–590

McIntosh T, Curran J (2009) Reducing semantic drift with bagging and distributional similarity. In: Proceedings of the Joint Conference of the 47th Annual Meeting of the ACL and the 4th International Joint Conference on Natural Language Processing of the AFNLP

Mur J (2008) Off-line answer extraction for question answering. PhD thesis, University of Groningen, Groningen

van Noord G (2004) Error mining for wide-coverage grammar engineering. In: Proceedings of the ACL 2004, Barcelona

van Noord G (2006) At last parsing is now operational. In: Mertens P, Fairon C, Dister A, Watrin P (eds) TALN06. Verbum Ex Machina. Actes de la 13e conference sur le traitement automatique des langues naturelles, pp 20–42

van Noord G (2009) Learning efficient parsing. In: Proceedings of the 12th Conference of the European Chapter of the Association for Computational Linguistics, Athens, Greece, pp 817–825

Padó S, Lapata M (2007) Dependency-based construction of semantic space models. Computational Linguistics 33(2):161–199

Pantel P, Pennacchiotti M (2006) Espresso: Leveraging generic patterns for automatically harvesting semantic relations. In: Proceedings of Conference on Computational Linguistics / Association for Computational Linguistics (COLING/ACL-06), Sydney, Australia, pp 113–120 van der Plas L (2008) Automatic lexico-semantic acquisition for question answering. PhD thesis, University of Groningen

Pollard C, Sag I (1994) Head-driven Phrase Structure Grammar. Center for the Study of Language and Information Stanford

Prins R, van Noord G (2001) Unsupervised pos-tagging improves parsing accuracy and parsing efficiency. In: IWPT 2001: International Workshop on Parsing Technologies, Beijing China

Ravichandran D, Hovy E (2002) Learning surface text patterns for a question answering system. In: Proceedings of ACL, vol 2, pp 41–47

Rinaldi F, Schneider G, Kaljurand K, Hess M, Romacker M (2006) An environment for relation mining over richly annotated corpora: the case of genia. BMC Bioinformatics 7

Rosario B, Hearst M (2004) Classifying semantic relations in bioscience texts. In: Proceedings of the 42nd Annual Meeting on Association for Computational Linguistics, Barcelona, Spain

Snow R, Jurafsky D, Ng A (2005) Learning syntactic patterns for automatic hypernym discovery. Advances in Neural Information Processing Systems 17:1297–1304

Soubbotin M, Soubbotin S (2002) Use of patterns for detection of answer strings: A systematic approach. In: Proceedings of TREC, vol 11

Stevenson M, Greenwood M (2009) Dependency pattern models for information extraction. Research on Language and Computation 3:13–39

Tiedemann J (2005) Integrating linguistic knowledge in passage retrieval for question answering. In: Proceedings of EMNLP 2005, Vancouver, pp 939–946

Tjong Kim Sang E, Bouma G, de Rijke M (2005) Developing offline strategies for answering medical questions. In: Mollá D, Vicedo JL (eds) AAAI 2005 workshop on Question Answering in Restricted Domains

Zhao S, Grishman R (2005) Extracting relations with integrated information using kernel methods. In: Proceedings of the 43rd Annual Meeting on Association for Computational Linguistics, Ann Arbor, Michigan, pp 419 – 426

Constraint-Satisfaction Inference for Entity Recognition

Sander Canisius, Antal van den Bosch and Walter Daelemans

Abstract One approach to QA answering is to match a question to candidate answers in a background corpus based on semantic overlap, possibly in combination with other levels of matching, such as lexical vector space similarity and syntactic similarity. While the computation of deep semantic similarity is as yet generally infeasible, semantic analysis in a specific domain is feasible, if the analysis is constrained to finding domain-specific entities and basic relations. Finding domain-specific entities, the focus of this chapter, is still not a trivial task due to ambiguities of terms. This problem, like many others in Natural Language Processing, is a sequence labelling task. We describe the development of a new approach to sequence labelling in general, based on the constraint satisfaction inference. The output of the machine-learning-based classifiers that solve aspects of the task (such as subsequently predicting the output of the label sequence) are considered as constraints on the global structured output analysis. The constraint-satisfaction inference method is compared to other state-of-the-art sequence labelling approaches, showing competitive performance.

1 Introduction

The Dutch word *arm* has two main senses, which in English are translated into *arm*, the body part, and *poor*. In the medical domain covered by the IMIX project, the *arm*

Sander Canisius
Netherlands Cancer Institute, Amsterdam, The Netherlands, e-mail: s.canisius@nki.nl

Antal van den Bosch
Tilburg center for Cognition and Communication, Tilburg University, Tilburg, The Netherlands, e-mail: Antal.vdnBosch@uvt.nl

Walter Daelemans
Computational Linguistics and Psycholinguistics Research Centre, University of Antwerp, Antwerp, Belgium, e-mail: Walter.Daelemans@ua.ac.be

meaning is obviously relevant: it is important for further processing, such as QA, to be able to detect when this sense is being used. The *poor* sense may be relevant to the domain as well, as it may be part of medical phrases such as *vitamin-poor diet*. Alternatively, it may be used in a non-medical sense in a medical text when it refers to the economic sense of *poor*. To summarise: detecting medically relevant concepts is not the mere detection of words from a gazetteer list (pre-collected lists of diseases, treatments, etc.), but rather a task that borrows some complexity from word sense disambiguation. Moreover, as many medical concepts are expressed in multiword expressions, such as *high fever* or *very low density lipoprotein*, there is also a challenge in finding the correct beginning and end of each expression.

Assuming that this aspect of automation requires a high-precision, high-recall solution to be used for the higher IMIX goal of medical question answering, we first define medical entity recognition as the focus problem, aiming to solve it with state-of-the-art natural language sequence processing methods. This chapter documents this particular effort. First, a limited set of thirteen entity types were identified that were relevant to the medical domain. Then, the IMIX medical encyclopaedia background corpus was annotated with these entities. A generic entity recognition method was developed, a hybrid of memory-based classification and constraint satisfaction inference, which was also applied to related benchmark problems in entity recognition as well as in purely syntactic chunking. The constraint satisfaction inference method is shown to attain state-of-the-art performance. However, entity recognition scores are not excellent. In this volume, Bouma, Fahmi, and Mur, in their chapter *Relation Extraction for Open and Closed Domain Question Answering*, further investigate the issue of how to employ the results of automatic entity recognition in QA.

The remainder of this chapter is structured as follows. Section 2 introduces sequence labelling as a special subclass of machine learning problems. In Section 3, a concise overview of previous machine learning approaches to sequence labelling is presented. Section 4 introduces a trigram-based sequence labelling method in which subsequences of trigram classes are predicted and simplistically resolved into output sequences. Next, Section 5 describes the constraint-satisfaction-based inference procedure. Experimental comparisons of a non-sequence-aware baseline classifier, the original trigram method, and the new classification and inference approach on a number of sequence labelling tasks are presented in Sections 6, 7 and 8, and discussed in Section 9. Conclusions are drawn in Section 10.

2 Sequence Labelling

Sequences are amongst the most versatile output structures in natural language processing. Many linguistic processing tasks can be seen as generating sequential outputs, either because the output naturally corresponds to a sequence, such as with POS-tagging, or because the targeted output structure is easily mapped to a sequence, as is the case in, for example, named-entity recognition. This chapter focuses on a subclass of sequence prediction, referred to as sequence labelling.

In a sequence labelling task, both inputs and outputs are sequences. Typically, the aim is not simply to classify the complete input sequence according to some global property, but rather to recover some type of hidden structure, closely linked to the elements of the input sequence. Sequence labelling abstractly defines this hidden structure as a sequence of label assignments to each of the elements of the input sequence. Thus, the output sequence—also referred to as the label sequence—has the same length as the input sequence, and there is a one-to-one correspondence between the elements of both sequences in the sense that the *i*th element of the output sequence is the label of the *i*th element of the input. Labels that make up such label sequences are taken from a restricted label set, and only have significance for the target application. In the context of sequence labelling, they are treated simply as atomic symbols. In this respect, a naive interpretation of sequence labelling would simply rephrase it as a sequence of multiclass classification cases. However, as in any structured prediction task, it is assumed that dependencies amongst different elements of the output sequence are as important as dependencies between an element of the input and its label in the output. In this chapter two *structured prediction* techniques are explored, one simple and one more complex. These techniques attempt to model dependencies between input and output elements as well as amongst elements of the output sequence.

3 Related Work

Because of the wide applicability of sequence labelling as a processing task template, it is unsurprising that it has received considerable attention in machine learning. Many techniques for learning to predict complex output structures were originally developed in the context of sequence labelling. This is the case, for example, for *structured linear models*, which in recent years have been the most popular framework for sequence labelling. There have been several different implementations of linear models for sequence labelling (Lafferty et al, 2001; Collins, 2002; Altun et al, 2003), which are based on the same graphical model formalism, but differ in the learning algorithm used for parameter estimation. The underlying graphical model encodes the independence assumption, which states that a certain label in the output sequence only depends on the input and a fixed number of labels directly preceding it in the output. This number is referred to as the Markov order of the model, and most commonly equals one, although second-order models have been used as well (e.g. Sha and Pereira, 2003). Because of this assumption, efficient inference is possible for such linear models. Most notably, the Viterbi algorithm finds the optimal output sequence according to a linear model in $O(L^2n)$ time, where L is the number of labels, and n is the length of the output sequence. On the negative side, features on output elements can only cover the small number of preceding labels permitted by the Markov order, and consequently models are restricted in the types of structural dependencies that can be modelled.

A range of other machine learning methods have been applied to sequence labelling as well. Punyakanok and Roth (2001) apply the constraint satisfaction with classifiers (CSCL) framework to sequence segmentation, formulated in terms of a sequence labelling problem. Both Ratnaparkhi (1996) and McCallum et al (2000) proposed discriminative Markov-like models for sequence labelling. They have mostly been superseded by structured linear models. As a final example in this incomplete overview, an output kernel approach to sequence labelling is described by Cortes et al (2005).

4 A Baseline Approach

Many tasks in natural language processing are sequence tasks, due to the obvious sequential nature of words as sequences of phonemes or letters, and sentences and spoken utterances as sequences of words. However, many machine learning methods do not typically learn these tasks by learning to map input sequences to output sequences. Rather, the standard approach that fits any supervised classification-based machine learning algorithm is to encode a sequence processing task by windowing, in which input subsequences are mapped to single output symbols. A single output symbol is typically associated with one of the input symbols, for example the middle one in the window.

Figure 1 displays this simplest version of the windowing process; fixed-width subsequences of input symbols are coupled to one output symbol. To ignore that the

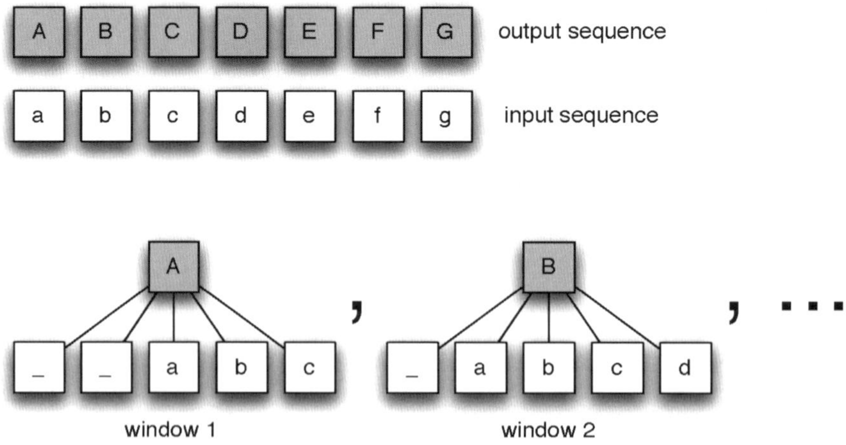

Fig. 1 Standard windowing process. Sequences of input symbols and output symbols are converted into windows of fixed-width input symbols each associated with one output symbol.

output forms a sequence is a problematic restriction, since it allows the classifier to produce invalid or impossible output sequences: for instance, it can make two neighbouring classifications in a sequence that are incompatible with each other, since it has no information about the other decision.

4.1 Class Trigrams

The restriction that classifications produce single output symbols is not intrinsic – the task may well be rephrased so that each input window is mapped to a sequence of output symbols. This directly prevents the classifier from predicting an invalid output sequence, since it will always produce sequences it has learned from training material. Van den Bosch and Daelemans (2006) propose to predict trigrams of labels (i.e. n-grams with $n = 3$) as a single atomic class label, thereby labelling three tokens at once.

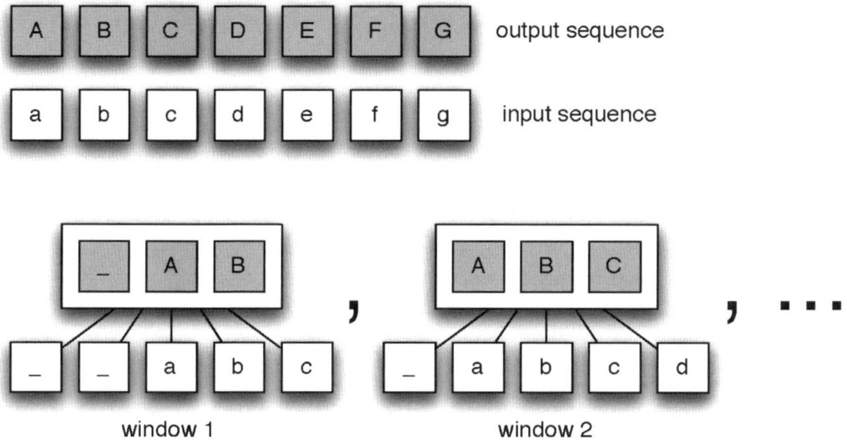

Fig. 2 Windowing process with n-grams of class symbols. Sequences of input symbols and output symbols are converted into windows of fixed-width input symbols each associated with, in this example, trigrams of output symbols.

Applying this general idea, Van den Bosch and Daelemans (2006) label each token with a complex class label composed of the labels for the preceding token, the token itself, and the one following it in the sequence. If such class trigrams are assigned to all tokens in a sequence, the actual label for each of those is effectively predicted three times, since every token but the first and last is covered by three class trigrams. Exploiting this redundancy, a token's possibly conflicting predictions are resolved by voting over them. If two out of three trigrams suggest the same label, this label is selected; in case of three different candidate labels, a classifier-specific confidence metric is used to break the tie.

Voting over class trigrams is but one possible approach to taking advantage of the redundancy obtained with predicting overlapping trigrams. A disadvantage of voting is that it discards one of the main benefits of the class trigram method: predicted class trigrams are guaranteed to be syntactically correct according to the training data. The voting technique splits up the predicted trigrams, and only refers to their unigram components when deciding on the output label for a token; no attempt is made to keep the trigram sequence intact in the final output sequence. The alternative to voting presented later in this chapter does attempt to retain predicted trigrams as part of the output sequence.

4.2 Memory-based Learning

The name memory-based learning refers to a class of methods based on the k-nearest neighbour rule. At training time, all example instances are stored in memory without attempting to induce an abstract representation of the concept to be learned. Generalisation is postponed until a test instance is classified. For a given test instance, the class predicted is the one observed most frequently among a number of most-similar instances in the instance base. By only generalising when confronted with the instance to be classified, a memory-based learner behaves as a local model, specifically suited for that part of the instance space that the test instance belongs to. In contrast, learners that abstract from their training data can only generalise globally. This distinguishing property makes memory-based learners especially suited for tasks where different parts of the instance space are structured according to different rules, as is often the case in natural-language processing.

For the experiments performed in this study the memory-based classifier was used as implemented in TiMBL[1] (Daelemans et al, 2009). In TiMBL, similarity is defined by two parameters: a feature-level similarity metric, which assigns a real-valued score to pairs of values for a given feature, and a set of feature weights, that express the importance of the various features for determining the similarity of two instances. To facilitate the explanation of the inference procedure in Section 5, this chapter will formally define some notions related to memory-based classification.

The function $N_{s,w,k}(x)$ maps a given instance x to the set of its nearest neighbours. Here, the parameters s, w, and k are the similarity metric, the feature weights, and the number k of nearest neighbours, respectively. They are considered as given in the following example, and this specific instantiation will therefore be referred to simply as $N(x)$. The function $w_d(c, N(x))$ returns the weight assigned to class c in the given neighbourhood according to the distance metric d; again the notation $w(c, N(x))$ is used to refer to a specific instantiation of this function. Using these two functions, the nearest neighbour rule can be formulated as follows.

---

[1] `http://ilk.uvt.nl/timbl`

$$\arg\max_{c} w(c, N(x))$$

The class c maximising the above expression is returned as the predicted class for the instance x.

5 Constraint Satisfaction Inference

One disadvantage of the voting method explained in Section 4.1 is that it ignores the fact that predicted class trigrams are guaranteed to be syntactically correct according to the training data. It is also blind to the overall quality of the output sequence it generates, as the voting is a local process. Both deficiencies are aimed to be repaired by adding constraint satisfaction inference as a global procedure for producing sequential output.

Constraints over an output space of label sequences are defined as where the constraints model relevant global dependencies in the predicted label sequence in order to apply constraint satisfaction inference to sequence labelling. In this case, these dependencies are implicitly encoded by the predicted trigrams. A strength of the class trigram method is the guarantee that any trigram that is predicted by the base classifier represents a syntactically valid subsequence of length three. This does not necessarily mean the trigram is a correct label assignment within the context of the current classification but it does reflect the fact that the trigram has been observed in the training data, and, moreover, is deemed most likely according to the base classifier's model. For this reason, it makes sense to try to retain as much as possible predicted trigrams in the output label sequence.

The inference method proposed in this section seeks to attain this goal by formulating the class trigram disambiguation task as a weighted constraint satisfaction problem (W-CSP). Constraint satisfaction is a well-studied research area with applications in numerous fields both inside and outside of computer science. Weighted constraint satisfaction extends the traditional constraint satisfaction framework with soft constraints; such constraints are not required to be satisfied for a solution to be valid, but constraints which satisfy a given solution, are rewarded according to weights assigned to them.

Formally, a W-CSP is a tuple (X, D, C, W). Here, $X = \{x_1, x_2, \ldots, x_n\}$ is a finite set of variables. $D(x)$ is a function that maps each variable to its domain, that is, the set of values that the variable can take on. C is the set of constraints. While a variable's domain dictates the values a single variable is allowed to take on, a constraint specifies which simultaneous value *combinations* over a number of variables are allowed. For a traditional (non-weighted) constraint satisfaction problem, a valid solution would be an assignment of values to the variables that (1) are a member of the corresponding variable's domain, and (2) satisfy *all* constraints in the set C. Weighted constraint satisfaction, however, relaxes this requirement to satisfy all constraints. Instead, constraints are assigned weights that may be interpreted as reflecting the importance of satisfying that constraint.

Let a constraint $c \in C$ be defined as a function that maps each variable assignment to 1 if the constraint is satisfied, or to 0 if it is not. In addition, let $W : C \rightarrow \mathbb{R}^+$ denote a function that maps each constraint to a positive real value, reflecting the weight of that constraint. Then, the optimal solution to a W-CSP is given by the following equation.

$$x^* = \arg\max_x \sum_c W(c)c(x)$$

That is, the assignment of values to its variables that maximises the sum of weights of the constraints that have been satisfied.

Translating the terminology introduced earlier in this chapter to the constraint satisfaction domain, each token of a sequence maps to a variable, the domain of which corresponds to the three candidate labels for this token suggested by the trigrams covering the token. This provides us with a definition of the function D, mapping variables to their domain. In the following, $y_{i,j}$ denotes the candidate label for token x_j predicted by the trigram assigned to token x_i.

$$D(x_i) = y_{i,i-1}, y_{i,i}, y_{i,i+1}$$

Constraints are extracted from the predicted trigrams. Given the goal of retaining predicted trigrams in the output label sequence as much as possible, the most important constraints are simply the trigrams themselves. A predicted trigram describes a subsequence of length three of the entire output sequence; turning such a trigram into a constraint is done with the intention of having this trigram end up in the final output sequence.

$$(x_{i-1}, x_i, x_{i+1}) = (y_{i,i-1}, y_{i,i}, y_{i,i+1}), \forall i$$

No base classifier is flawless though, and therefore not all predicted trigrams can be expected to be correct. Nevertheless, even an incorrect trigram may carry some useful information regarding the output sequence: one trigram also covers two bigrams, and three unigrams. An incorrect trigram may still contain smaller subsequences, of length one or two, that are correct. Therefore, all of these are also mapped to constraints.

$$(x_{i-1}, x_i) = (y_{i,i-1}, y_{i,i}), \qquad \forall i$$
$$(x_i, x_{i+1}) = (y_{i,i}, y_{i,i+1}), \qquad \forall i$$

$$x_{i-1} = y_{i,i-1}, \qquad \forall i$$
$$x_i = y_{i,i}, \qquad \forall i$$
$$x_{i+1} = y_{i,i+1}, \qquad \forall i$$

With such an amount of overlapping constraints, the satisfaction problem obtained easily becomes over-constrained. This means no variable assignment exists that can satisfy all constraints without breaking another. Only one incorrectly predicted class trigram already leads to two conflicting candidate labels for one of the tokens at least. Yet, without conflicting candidate labels no inference would be needed to start with. The choice for the weighted constraint satisfaction method always allows a solution to be found, even in the presence of conflicting constraints. Rather than requiring all constraints to be satisfied, each constraint is assigned a certain weight; the optimal solution to the problem is an assignment of values to the variables that optimise the sum of the weights of the constraints that are satisfied.

Constraints can be directly traced back to a prediction made by the base classifier. If two constraints are in conflict, the one which the classifier was most certain of should preferably be satisfied. In the W-CSP framework, this preference can be expressed by weighting constraints according to the classifier confidence for the originating trigram. For the memory-based learner, the confidence of the classifier is for a predicted class c_i is defined as the weight assigned to that class in the neighbourhood of the test instance, divided by the total weight of all classes.

$$\frac{w(c_i, N(x))}{\sum_{c} w(c, N(x))}$$

Let x denote a test instance, and c^* its predicted class. Constraints derived from this class are weighted according to the following rules.

- For a trigram constraint, the weight is the base classifier's confidence value for the class c^*;
- For a bigram constraint, the weight is the sum of the confidences for all trigram classes in the nearest-neighbour set of x that assign the same label bigram to the tokens spanned by the constraint;
- For a unigram constraint, the weight is the sum of the confidences for all trigram classes in the nearest-neighbour set of x that assign the same label to the token spanned by the constraint.

5.1 Solving the CSP

The output space of sequence labelling tasks is exponential in the length of the input sequence. Since the constraint satisfaction problem is constructed to only consider solutions that can be built from predictions made by the base classifiers, the output space searched by the constraint solver will typically be substantially smaller than this full output space. In spite of that, though, the worst-case complexity of this solution space remains exponential, be it with a smaller base. The constraint solver in the sequence labelling approach has to search this space.

This issue is, however, not unique to constraint satisfaction inference. It is faced by any inference-based sequence labelling approach. This is why solutions for inference for sequence labelling already exist. A popular approach is to use the Viterbi algorithm. Under the Markov assumption, finding the optimal solution is guaranteed. On the downside, the Markov assumption restricts the dependencies that are modelled. As a less restrictive alternative to the Viterbi algorithm, approximate search algorithms have been employed, such as beam search (Ratnaparkhi, 1996), or simulated annealing (Finkel et al, 2005). Viterbi is by far the most popular inference algorithm, and it could also serve as the basis of the constraint solver. However, the method allows for modelling unrestricted dependencies, not just the type of dependencies meeting the Markov assumption. This choice leaves only exhaustive or approximate search as possibilities. For the experiments in this chapter, an exhaustive search has been chosen. Even though, in the worst case, this leads to exponential-time inference, the search space resulting from the CSP formulation is assumed to be sufficiently small to make exhaustive search feasible.

6 Sequence Labelling Tasks

As mentioned earlier, natural language processing offers a wide array of tasks that can be seen as instances of sequence labelling. Input sequences may be words, sentences, or even documents. The output sequences correspond to a direct one-to-one labelling of the elements of the input sequence, or may for example encode a segmentation of the input sequence. While sequence segmentation tasks may be performed with special-purpose approaches (Carreras, 2005; Sarawagi and Cohen, 2005; Daumé III, 2006), they are most often reformulated as per-token sequence labelling tasks using the encoding scheme proposed by Ramshaw and Marcus (1995), generally referred to as the `IOB` encoding. According to this encoding, symbols assigned to input tokens signal whether the token is inside a segment (`I`), outside a segment (`O`), or at the start of a segment that was preceded by a segment of the same type ("between", `B`). A variant of `IOB` sometimes referred to as `BIO` uses the `B` to mark all tokens at the beginning of segments. Both `IOB` and `BIO` are used in the experiments reported below. Segment type labels are appended to this symbol to denote the current type of segment; `B-PP` which would mark the beginning of a prepositional phrase segment in syntactic chunking.

To illustrate the applicability of constraint satisfaction inference for a range of tasks, the results on three processing tasks are presented that can all be approached as sequence labelling tasks. All three tasks take the sentence as their domain, operating on sequences of word tokens; one is a syntactic task, and the remaining two are named entity recognition tasks. One of these entity recognition tasks is of direct interest to the IMIX project, namely the identification of medical entities in Dutch medical encyclopaedic text. The three tasks are introduced briefly in the following subsections.

6.1 Syntactic Chunking

In syntactic chunking, sometimes referred to as shallow parsing, the goal is to divide an input sentence into non-recursive syntactic base phrases, or chunks. Each base phrase is centred around some head word, which gives rise to the syntactic type of the phrase, and in addition includes some of the modifying words of the head. For example, a noun phrase consists of a head noun, and may also include some adjectives and determiners directly preceding, and syntactically modifying, the head word. As part of the chunking task, the exact boundaries of each chunk have to be determined, and in addition each chunk has to be labelled with its syntactic type. Figure 3 shows an example of a sentence and the base phrases that are to be recognised as part of the syntactic chunking task.

[$_{\mathbf{NP}}$ Rockwell International Corp.] [$_{\mathbf{NP}}$'s Tulsa unit] [$_{\mathbf{VP}}$ said] [$_{\mathbf{NP}}$ it] [$_{\mathbf{VP}}$ signed] [$_{\mathbf{NP}}$ a tentative agreement] [$_{\mathbf{VP}}$ extending] [$_{\mathbf{NP}}$ its contract] [$_{\mathbf{PP}}$ with] [$_{\mathbf{NP}}$ Boeing Co.] [$_{\mathbf{VP}}$ to provide] [$_{\mathbf{NP}}$ structural parts] [$_{\mathbf{PP}}$ for] [$_{\mathbf{NP}}$ Boeing] [$_{\mathbf{NP}}$'s 747 jetliners] .

Fig. 3 Example sentence from the syntactic chunking task.

The standard benchmark for syntactic chunking is the data set created for the CoNLL-2000 shared task (Tjong Kim Sang and Buchholz, 2000). For this data set, chunks have been extracted from the full syntactic annotation of the Wall Street Journal part of the Penn Treebank (Marcus et al, 1993) and encoded in the aforementioned `BIO` notation. Besides the words and the chunks for each sentence, POS tags for those words have been obtained by running the POS tagger by Brill (1994) on the data. Using these predicted POS tags instead of the manually assigned tags available in the original corpus makes for a more realistic scenario, where the chunker has to cope with tagging errors in its input. The training data of the CoNLL-2000 data set corresponds to sections 15 to 18 of the WSJ corpus, while section 20 serves as test data. In addition, section 21 is frequently used as a development set, mainly for tuning learning algorithm parameters. This same train-test-development split has been used for the experiments reported in this chapter.

The features used for this task are all fairly standard. In a window of five tokens centred around the focus token, there are features for the word form, POS tag, and a symbol encoding certain orthographical features of the word. In addition, in a window of three tokens centred around the focus token, conjunctions of pairs of consecutive words are included, and the same for their POS tags. Finally, again in a three-token window, conjunctions of word forms and their POS tag are encoded as features. Figure 4 illustrates these features in the context of an example sentence.

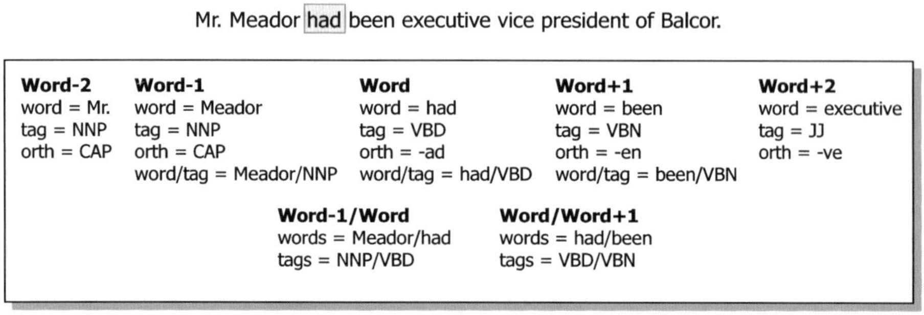

Fig. 4 Feature representation for the word "had" in the above sentence as used with the syntactic chunking task.

6.2 Named-Entity Recognition

An important subtask of information extraction is identifying names in running text, and characterising the type of real-world entity they refer to. Named-entity recognition, as it is called, has been the subject of several organised evaluations, such as MUC (Chinchor, 1995), CoNLL (Tjong Kim Sang, 2002; Tjong Kim Sang and De Meulder, 2003), BioCreative (Hirschman et al, 2005), and ACE (Doddington et al, 2004). Each of these evaluations provided annotated data sets for several types of entities, and participating systems had to learn how to recognise them. Interestingly, the notion of a named entity can be defined as broadly or narrowly, as is suited for the task at hand. For example, in broadcast news texts, persons and organisations are relevant entities to discover. In contrast, those entities may not be worthwhile at all in biomedical texts, where references to proteins and viruses are more likely to be of interest.

The named-entity recognition task considered in this chapter has been defined as part of the CoNLL-2003 shared task (Tjong Kim Sang and De Meulder, 2003), from which the English data is used. Texts from this data set have been collected from the Reuters corpus (Lewis et al, 2004), which consists of general news stories, and have been annotated for named-entity mentions using the aforementioned `IOB` notation. The named entities to be recognised have been divided in four classes: people, locations, organisations, and a rather broad miscellaneous category, including such entity types as languages, events, and book titles. The task involves both identifying the exact boundaries of each named-entity mention and assigning the correct entity type. Following the standard partitioning of the shared task data, the "test A" subset for development purposes was used, and the "test B" subset for the final evaluation.

[$_{\mathbf{ORG}}$ Interior] Minister [$_{\mathbf{PER}}$ Zbigniew Siemiatkowski] and [$_{\mathbf{PER}}$ Bernd Schmidbauer] , [$_{\mathbf{MISC}}$ German] intelligence co-ordinator in [$_{\mathbf{PER}}$ Helmut Kohl] 's chancellery, sealed the closer links during talks in [$_{\mathbf{LOC}}$ Warsaw] .

Fig. 5 Example sentence from the named-entity recognition task.

It is insightful to compare the named-entity recognition and syntactic chunking tasks. Both are sentence-level segmentation and labelling tasks. However, whereas in syntactic chunking, tokens that are not part of a chunk are exceptions, in named-entity recognition, the vast majority of tokens will not be part of a named-entity segment. For this reason, sequential correlation within label sequences can be expected to be different. The fact that a noun phrase is likely to be followed by a verb phrase can be a valuable clue to a learner. In contrast, observing that a named-entity is often followed by tokens that do not refer to an entity is almost stating the obvious. Nevertheless, sequential correlation is still an important factor in named-entity recognition as well. It may help in deciding that the two-token phrase *George Washington* is more likely to be a single `Person` entity, than a `Person` entity followed by a `Location`.

The feature set used for named-entity recognition includes features similar to those used for syntactic chunking, as displayed in Figure 4. It is extended with some additional features. First, features encoding affixes of lengths 2 and 3 of the word in focus are included. Second, syntactic chunk tags for the words in a three-token window are included. Third, again in a three-token window, features signalling the presence of the word in entity-specific gazetteer lists were added. The gazetteer lists used for this purpose were provided as part of the CoNLL-2003 shared task, and merely list the entities that are found in the annotated training data.

6.3 Medical Concept Chunking: The IMIX Task

The IMIX data set is yielded from a manually annotated Dutch-language medical encyclopaedia. The annotation offers labels for various medical entities, such as disease names, body parts, and treatments, forming a set of twelve entity types in total. Table 1 lists the twelve entity types and their frequency of occurrence in the encyclopaedia, as well as their average numbers of tokens. Three of the twelve entity types, *disease symptom*, *duration*, and *advice*, are typically several tokens long, which is to be expected given their more circumscriptive nature. The types *person feature* (such as gender, weight, age range), *treatment*, and *method of diagnosis* often span two tokens. The remaining six types are mostly single-token entities. The data has been split into training and test sets, resulting in 428,502 training examples (i.e., tokens, of which 128,182 tokens or 29.9% belong to a medical entity), and 47,430 test examples (with 14,314 or 39.8% of the tokens beloning to a medical entity).

Table 1 The twelve medical entity types in the IMIX data, their frequency of occurrence in the training data, and the average number of tokens per entity.

Entity type	Number of entities	Av. tokens per entity
Body part	21,509	1.2
Disease	16,037	1.3
Treatment	7,210	1.6
Disease symptom	6,636	2.2
Person	5,052	1.3
Bodily function	3,453	1.3
Duration	2,563	3.7
Disease feature	2,004	1.2
Microorganism	1,672	1.3
Method of diagnosis	1,447	1.5
Person feature	1,091	1.7
Advice	168	5.7

The feature set used for the domain-specific medical named-entity recognition is a simple encoding of the local context of each word. It consists of a seven-token window of words and POS tags centred on the token for which the label is predicted.

Een [$_{\textbf{disease}}$ cefaalhematoom] is een [$_{\textbf{disease}}$ bloeduitstorting] onder de [$_{\textbf{body-part}}$ hoofdhuid] . Bij [$_{\textbf{disease}}$ infantiel botulisme] kunnen in extreme gevallen [$_{\textbf{symptom}}$ ademhalingsproblemen] en [$_{\textbf{symptom}}$ algehele lusteloosheid] optreden.

Fig. 6 Example sentences from the IMIX concept chunking task.

7 Experimental Set-up

7.1 Evaluation

Arguably the purest way to evaluate sequence labelling performance is to measure the proportion of complete label sequences that are predicted correctly. This is an extremely strict measure that ignores the fact that even partly incorrect label sequences may still be usable in many applications. A more forgiving performance criterion counts the proportion of individual labels predicted correctly. As an advantage of token accuracy, label sequences that are almost, but not completely predicted correctly still contribute proportionally to the performance score. For some tasks—such as POS-tagging, which is not dealt with in this chapter—token accuracy is the most natural evaluation measure. However, for many other tasks in natural language processing, token accuracy is rather uninformative. A typical example in this respect is named-entity recognition when approached as an `IOB` labelling task. The vast majority of tokens in the correct output have the `O` label; therefore, a classifier that always predicts `O` is likely to attain high token accuracy. However, in named-entity recognition, one is not interested in the tokens outside of named-entity segments, but only in those inside them. Token accuracy assigns too little importance to those tokens, and is therefore mostly unsuited for named-entity recognition.

The three tasks reported on in this chapter are in fact also concerned with sequence segmentation, rather than pure sequence labelling. Both sequence accuracy and token accuracy do not specifically measure the quality of the segments found. The most common metrics that do specifically measure segmentation and optional labelling quality are precision, recall, and $F_{\beta=1}$ (Van Rijsbergen, 1979). *Precision* corresponds to the proportion of predicted segments that are correct, where a segment will be considered correct if both its boundaries and its label exactly matches the true segment. *Recall* measures the proportion of true segments that have indeed been predicted correctly. $F_{\beta=1}$, finally, matches the harmonic mean of precision and recall.

7.2 Constraint Prediction

The choice for n-gram constraints still leaves the parameter n to be tuned, or at the very least, to be fixed in advance. Indeed, n can be seen as a genuine parameter of constraint satisfaction inference for sequence labelling; one that can be tuned for every new sequence labelling task that is to be performed. In this chapter, it was chosen not to do this, but instead decide on an $n = 3$ for all three tasks. Consequences of a choice of n include the following.

- The higher the value of n, the sparser the training data for the constraint predictor will be. In the extreme case, the label to be predicted matches the entire label sequence. This is generally not a viable option.
- n is also the theoretically maximal size of the micro-label domains. For a sequence of length T, this implies that the size of the worst-case output space is in the order of $O(n^T)$. Consequently, high values for n inevitably require approximate search in the output space. Gains that might potentially result from high values for n, may be lost as a result of only being able to use approximate search.
- Larger n-gram constraints also result in an increase of the number of, possibly conflicting, constraints covering a single micro-label. Although weighting the constraints by classification confidence should ensure that correct constraints are satisfied, too many incorrect constraints that conflict with the correct ones, may still overrule the latter.

For the above reasons, n is preferred not to be too high. On the other hand, choosing n too small will result in the loss of valuable structural information. Although there may certainly exist sequence labelling tasks for which 5-gram constraint satisfaction inference will be a viable option, or will even lead to better performance than trigrams, trigram constraints are chosen as the basis for all experiments in this chapter. An attractive consequence of choosing trigram constraints is that the solution space yielded by them is rather small, and therefore allows for efficient inference. Finally, as an interesting parallel with Markov-based sequence labelling approaches, trigram constraints can be seen as modelling an undirected first-order Markov assumption, where a label depends on the labels preceding and following it.

8 Results

For the experiments, memory-based learners were trained and automatically optimised with wrapped progressive sampling (Van den Bosch, 2004) to predict class trigrams for each of the three tasks introduced above. Table 2 lists the performances of constraint satisfaction inference compared to the majority voting method, both applied to the output of the base classifiers, and compares them with the performance of a naive baseline method that treats each token as a separate classification case without coordinating decisions over multiple tokens.

Table 2 Performances (F-scores) of the baseline method, and the trigram method combined both with majority voting, and with constraint satisfaction inference. The last column shows the performance of the hypothetical oracle inference procedure.

Task	Baseline	Voting	CSI	Oracle
CHUNK	91.6	93.1	**93.8**	96.2
NER	76.5	82.5	**85.6**	88.9
IMIX	64.7	67.5	**68.9**	74.9

The column labelled "Oracle" in Table 2 represents the performance of the constraint-satisfaction inference method if it would be able, through an oracle, to choose the correct base classifier output among conflicting outputs. The "Oracle" scores thus represent an upper-bound score where all remaining errors are due to the base classifier.

The results show that both the majority voting method and constraint satisfaction inference outperform the naive baseline classifier. In turn, constraint satisfaction inference outperforms majority voting consistently. This shows that, given the same sequence of predicted trigrams, the global constraint satisfaction inference manages better to recover sequential correlation than majority voting. On the other hand, the error reduction attained by majority voting with respect to the naive baseline is more impressive than the one obtained by constraint satisfaction inference with respect to majority voting. However, it should be emphasised that, while both methods trace back their origins to the work of Van den Bosch and Daelemans (2006), constraint satisfaction inference is not applied after, but instead of majority voting. This means that the error reduction attained by majority voting is also attained, independently by constraint satisfaction inference, but in addition constraint satisfaction inference manages to improve performance on top of that.

8.1 Comparison to Alternative Techniques

In the experiments reported on, constraint satisfaction inference has been compared with a naive baseline and the majority-voting-based sequence labelling technique in a set-up where the feature sets were the same for all techniques. Although this is arguably the most objective approach to such a comparison, it is certainly true that some methods might actually perform better with more or different types of features. For this reason, another interesting comparison is with other published work using the same data sets. The data sets for syntactic chunking and named-entity recognition have both been standardised as part of the CoNLL shared task, and consequently many additional results on those data sets are available. The top-performing systems for both tasks will be briefly discussed.

Table 3 Comparison of the performance of constraint satisfaction inference on syntactic chunking with other published results.

	Prec	Rec	$F_{\beta=1}$
Ando and Zhang (2005)	94.57	94.20	94.39
Zhang et al (2002)	94.28	94.07	94.17
Kudo and Matsumoto (2001)	93.89	93.92	93.91
CSI	**93.81**	**93.80**	**93.80**
Carreras (2005)	94.20	93.38	93.79
Kudo and Matsumoto (2000)	93.45	93.51	93.48

First of all, the top-performing systems for syntactic chunking are listed in Table 3. The CoNLL-2000 shared task on syntactic chunking took place before machine learning techniques for structured prediction gained widespread popularity. The best system at the time (Kudo and Matsumoto, 2000), using a recurrent sliding-window approach and an extensive set of features, attained an F-score of 93.48. Most other systems scored considerably lower. Structured prediction approaches published since then easily outperform those scores, as does the approach taken here. In fact, most such approaches attain similar scores, as can be seen when looking at the scores for Carreras (2005), Kudo and Matsumoto (2001), and constraint satisfaction inference. Two systems perform substantially better, though in both cases, this performance gain can be attributed to additional information sources. Zhang et al (2002) use an enriched feature set that includes the output of a full syntactic parser. Without those features, their system reaches an F-score of 93.57. Ando and Zhang (2005) manage to improve performance by employing semi-supervised learning. Their fully supervised system achieves a performance of 93.60.

For named-entity recognition, the picture is slightly different. As can be seen in Table 4, the top-performing systems in the CoNLL-2003 shared task are still amongst the best published to date. These systems did not put extensive effort in structured prediction approaches, but rather used abundant sets of features. Possibly, good global coordination is not as essential for named-entity recognition as it is for syntactic chunking. The fact that most tokens that are part of entities belong to a single class allows carefully crafted local classifiers to perform at a high level for this task. In that light, constraint satisfaction performs rather well, given the limited effort invested in feature optimisation. As an illustration of the complexity of the top-performing systems, Florian et al (2003) combine four different classifiers, and use gazetteer lists comprising tens of thousands of words and even integrate the output of two other entity classifiers. Chieu and Ng (2003) also used large gazetteer lists and in addition, performed extensive feature engineering. Finally, as with syntactic chunking, the best scores published so far involve semi-supervised learning (Ando and Zhang, 2005).

Table 4 Comparison of the performance of constraint satisfaction inference on named-entity recognition with other published results.

	Prec	Rec	$F_{\beta=1}$
Ando and Zhang (2005)	-	-	89.31
Florian et al (2003)	88.99	88.54	88.76
Chieu and Ng (2003)	88.12	88.51	88.31
Klein et al (2003)	86.12	86.49	86.31
CSI	**85.88**	**85.29**	**85.58**
Zhang and Johnson (2003)	86.13	84.88	85.50
Carreras et al (2003)	84.05	85.96	85.00

9 Discussion

The experiments reported on in the previous section showed that, by globally evaluating the quality of possible output sequences, the constraint satisfaction inference procedure manages to attain better results than the original majority-voting approach. This section attempts to further analyse the behaviour of the inference procedure.

There is a subtle balance between the quality of the trigram-predicting base classifier, and the gain that any inference procedure for trigram classes can reach. If the base classifier's predictions are perfect, all three candidate labels will agree for all tokens in the sequence; consequently the inference procedure can only choose from one potential output sequence. At the other extreme, if all three candidate labels disagree for all tokens in the sequence, the inference procedure's task is to select the best sequence among 3^n possible sequences, where n denotes the length of the sequence; it is unlikely that such a huge number of candidate label sequences could be dealt with appropriately.

Table 5 collects the base classifier accuracies, and the average number of potential output sequences per sentence resulting from its predictions. For all tasks, the number of potential sequences is manageable; far from the theoretical maximum 3^n. This is an important observation, since it shows that the output space spanned by the predicted class trigrams is small enough to be searched exhaustively, which in fact was done for all three tasks. Exhaustive search, rather than for example Viterbi search, allows for constraints to cover arbitrary parts of the complete output label sequence. Although this feature was not employed in the current experiments, modelling of higher-level structural dependencies could prove beneficial in optimising entity recognition performance.

Table 5 The average number of potential output sequences that result from class trigram predictions made by a memory-based base classifier.

Task	Base acc.	Avg. # seq.
CHUNK	88.2	57.8
NER	92.3	19.1
IMIX	77.1	9.3

9.1 Other Constraint-based Approaches to Sequence Labelling

Using constraint satisfaction for global optimisation in sequence learning has been explored by others as well. As the following brief overview shows, constraints in these approaches do not stem from base classifiers, but are based on external knowledge.

Constraint Satisfaction with Classifiers (Punyakanok and Roth, 2001) performs the somewhat more specific task of identifying phrases in a sequence. Like the method described here, the task of coordinating local classifier decisions is formulated as a constraint satisfaction problem. The variables encode whether or not a certain contiguous span of tokens forms a phrase. Hard constraints enforce that no two phrases in a solution overlap.

In a similar way to this method, classifier confidence estimates are used to rank solutions in order of preference. Unlike in this method, however, both the domains of the variables and the constraints are prespecified; the classifier is used only to estimate the cost of potential variable assignments. In this approach, the classifier predicts the domains of the variables, the constraints, and the weights of those.

Roth and Yih (2005) replace the Viterbi algorithm for inference in conditional random fields with an integer linear programming formulation. This allows arbitrary global constraints to be incorporated in the inference procedure. Essentially, the method adds constraint satisfaction functionality on top of the inference procedure. In this method, constraint satisfaction *is* the inference procedure. Nevertheless, arbitrary global constraints (both hard and soft) can easily be incorporated in this framework as well.

10 Conclusion

The classification and inference approach is a popular and effective framework for performing sequence labelling in tasks where there is strong interaction between output labels. Most existing approaches to sequence labelling use a base classifier as part of a scoring function to guide the search through the output space of all possible label sequences. The constraint satisfaction inference approach presented in this chapter is different in the sense that the base classifier predictions are not only used to score candidate label sequences, but also to restrict the solution space that is explored during inference.

Constraint satisfaction inference builds on the class trigram method introduced by Van den Bosch and Daelemans (2006), but reinterprets it as a strategy for generating multiple potential output sequences, from which it selects the sequence that has been found to be most optimal according to a weighted constraint satisfaction formulation of the inference process. In a series of experiments involving three sequence labelling tasks, covering both syntactic and semantic processing, constraint satisfaction inference has been shown to improve substantially on the performance achieved by a simpler inference procedure based on majority voting, which was proposed the original work on the class trigram method. In addition, the method was found to perform on a par with competing sequence labelling methods.

The work presented in this chapter shows there is potential for alternative interpretations of the classification and inference framework. Advantages of this method may be found in faster inference, no restrictions on the type of dependencies that can be modelled, and the fact that it can be used in combination with any type of base classifier. Sequence labelling, while an important class of machine learning problems in natural language processing, is not the only structured output domain to which the constraint satisfaction inference method can be applied. To illustrate this, the same method was applied to syntactic parsing Canisius and Tjong Kim Sang (2007) and machine translation Canisius and Van den Bosch (2009).

In the larger framework of domain-specific QA, a proper recognition of domain-specific entities in background material with high precision and high recall is important for achieving improved results. The work of Bouma, Fahmi, and Mur (Chapter 9, this volume) complements the work by taking on the empirical question to whether domain-specific entity recognition aids QA.

References

Altun Y, Tsochantaridis I, Hofmann T (2003) Hidden markov support vector machines. In: Fawcett T, Mishra N (eds) Proceedings of the Twentieth International Conference on Machine Learning (ICML 2003), pp 3–10

Ando R, Zhang T (2005) A high-performance semi-supervised learning method for text chunking. In: Proceedings of the 43rd Annual Meeting of the Association for Computational Linguistics, pp 1–9

Brill E (1994) Some advances in transformation-based part-of-speech tagging. In: Proceedings AAAI '94

Canisius S, Tjong Kim Sang E (2007) A constraint satisfaction approach to dependency parsing. In: Proc. of the CoNLL Shared Task Session of EMNLP-CoNLL 2007, Prague, Czech Republic, pp 1124–1128

Canisius S, Van den Bosch A (2009) A constraint satisfaction approach to machine translation. In: Proceedings of the 13th Annual Conference of the European Association for Machine Translation (EAMT-2009), pp 182–189

Carreras X (2005) Learning and inference in phrase recognition: A filtering-ranking architecture using perceptron. PhD thesis, Universitat Politècnica de Catalunya

Carreras X, Màrquez L, Padró L (2003) A simple named entity extractor using adaboost. In: Daelemans W, Osborne M (eds) Proceedings of the seventh Conference on Natural Language Learning at HLT-NAACL 2003, pp 152–155

Chieu H, Ng H (2003) Named entity recognition with a maximum entropy approach. In: Daelemans W, Osborne M (eds) Proceedings of the seventh Conference on Natural Language Learning at HLT-NAACL 2003, pp 160–163

Chinchor N (1995) Named entity task definition. In: Proceedings of the Sixth Message Understanding Conference (MUC-6), pp 317–332

Collins M (2002) Discriminative training methods for hidden Markov models: theory and experiments with perceptron algorithms. In: Hajic J, Matsumoto Y (eds) Proceedings of the ACL-02 conference on Empirical Methods in Natural Language Processing, pp 1–8

Cortes C, Mohri M, Weston J (2005) A general regression technique for learning transductions. In: Raedt LD, Wrobel S (eds) Proceedings of the Twenty-Second International Conference on Machine Learning (ICML 2005), pp 153–160

Daelemans W, Zavrel J, Van der Sloot K, Van den Bosch A (2009) TiMBL: Tilburg memory based learner, version 6.2, reference guide. Tech. Rep. ILK 09-01, ILK Research Group, Tilburg University

Daumé III H (2006) Practical structured learning techniques for natural language processing. PhD thesis, University of Southern California

Doddington G, Mitchell A, Przybocki M, Ramshaw L, Strassel S, Weischedel R (2004) The Automatic Content Extraction (ACE) Program–Tasks, Data, and Evaluation. In: Proceedings of the Fourth International Conference on Language Resources and Evaluation (LREC'04), pp 837–840

Finkel J, Grenager T, Manning C (2005) Incorporating non-local information into information extraction systems by Gibbs sampling. In: Proceedings of the 43rd Annual Meeting of the Association for Computational Linguistics (ACL'05), pp 363–370

Florian R, Ittycheriah A, Jing H, Zhang T (2003) Named entity recognition through classifier combination. In: Daelemans W, Osborne M (eds) Proceedings of the seventh Conference on Natural Language Learning at HLT-NAACL 2003, pp 168–171

Hirschman L, Yeh A, Blaschke C, Valencia A (2005) Overview of BioCreAtIvE: critical assessment of information extraction for biology. BMC Bioinformatics 6(S1)

Klein D, Smarr J, Nguyen H, Manning C (2003) Named entity recognition with character-level models. In: Daelemans W, Osborne M (eds) Proceedings of the seventh Conference on Natural Language Learning at HLT-NAACL 2003, pp 180–183

Kudo T, Matsumoto Y (2000) Use of support vector learning for chunk identification. In: Cardie C, Daelemans W, Nedellec C, Tjong Kim Sang E (eds) Proceedings of the Fourth Conference on Computational Natural Language Learning and of the Second Learning Language in Logic Workshop, pp 142–144

Kudo T, Matsumoto Y (2001) Chunking with support vector machines. In: Second meeting of the North American Chapter of the Association for Computational Linguistics on Language technologies, pp 1–8

Lafferty J, McCallum A, Pereira F (2001) Conditional random fields: Probabilistic models for segmenting and labeling sequence data. In: Proceedings of the 18th International Conference on Machine Learning, Williamstown, MA

Lewis D, Yang Y, Rose T, Li F (2004) RCV1: A New Benchmark Collection for Text Categorization Research. Journal of Machine Learning Research 5:361–397

Marcus M, Santorini S, Marcinkiewicz M (1993) Building a Large Annotated Corpus of English: the Penn Treebank. Computational Linguistics 19(2):313–330

McCallum A, Freitag D, Pereira F (2000) Maximum entropy Markov models for information extraction and segmentation. In: Langley P (ed) Proceedings of the Seventeenth International Conference on Machine Learning (ICML 2000), pp 591–598

Punyakanok V, Roth D (2001) The use of classifiers in sequential inference. In: NIPS-13; The 2000 Conference on Advances in Neural Information Processing Systems, The MIT Press, pp 995–1001

Ramshaw L, Marcus M (1995) Text chunking using transformation-based learning. In: Proceedings of the 3rd ACL/SIGDAT Workshop on Very Large Corpora, Cambridge, Massachusetts, USA, pp 82–94

Ratnaparkhi A (1996) A maximum entropy part-of-speech tagger. In: Proceedings of the Conference on Empirical Methods in Natural Language Processing, May 17-18, 1996, University of Pennsylvania

Roth D, Yih W (2005) Integer linear programming inference for conditional random fields. In: Proceedings of the 22nd International Conference on Machine Learning, ACM, p 743

Sarawagi S, Cohen W (2005) Semi-markov conditional random fields for information extraction. In: Advances in Neural Information Processing Systems, vol 17, pp 1185–1192

Sha F, Pereira F (2003) Shallow parsing with conditional random fields. In: Proceedings of the 2003 Conference of the North American Chapter of the Association for Computational Linguistics on Human Language Technology, pp 134–141

Tjong Kim Sang E (2002) Introduction to the conll-2002 shared task: Language-independent named entity recognition. In: Proceedings of CoNLL-2002, Taipei, Taiwan, pp 155–158

Tjong Kim Sang E, Buchholz S (2000) Introduction to the CoNLL-2000 shared task: Chunking. In: Proceedings of CoNLL-2000 and LLL-2000, pp 127–132

Tjong Kim Sang E, De Meulder F (2003) Introduction to the CoNLL-2003 shared task: Language-independent named entity recognition. In: Daelemans W, Osborne M (eds) Proceedings of CoNLL-2003, Edmonton, Canada, pp 142–147

Van den Bosch A (2004) Wrapped progressive sampling search for optimizing learning algorithm parameters. In: Verbrugge R, Taatgen N, Schomaker L (eds) Proceedings of the Sixteenth Belgian-Dutch Conference on Artificial Intelligence, Groningen, The Netherlands, pp 219–226

Van den Bosch A, Daelemans W (2006) Improving sequence segmentation learning by predicting trigrams. In: Proceedings of the Ninth Conference on Natural Language Learning, CoNLL-2005, Ann Arbor, MI, pp 80–87

Van Rijsbergen C (1979) Information Retrieval. Buttersworth, London

Zhang T, Johnson D (2003) A robust risk minimization based named entity recognition system. In: Daelemans W, Osborne M (eds) Proceedings of the seventh Conference on Natural Language Learning at HLT-NAACL 2003, pp 204–207

Zhang T, Damerau F, Johnson D (2002) Text chunking based on a generalization of winnow. Journal of Machine Learning Research 2:615–637

Extraction of Hypernymy Information from Text*

Erik Tjong Kim Sang, Katja Hofmann and Maarten de Rijke

Abstract This chapter presents the results of three studies in extracting hypernymy information from a text. In the first, a method based on a single extraction pattern applied to the web is compared with a set of patterns applied to a big corpus. In the second study, it is examined how relation extraction can be performed reliably from a text without having access to a word sense tagger. And in a third experiment, it is checked what the effect of elaborate syntactic information has on the extraction process. Both using more data and the removal of ambiguities from the training data is found to be beneficial for the extraction process. But it is surprising to find a positive effect of additional syntactic information.

1 Introduction

Lexical taxonomies, such as WordNet, are important resources underlying natural language processing techniques such as machine translation and word-sense disambiguation. However, manual creation and maintenance of such taxonomies are tedious and time-consuming tasks. This has led to a great deal of interest in automatic methods for retrieving taxonomic relations. Efforts for both manual development of taxonomies and automatic acquisition methods have largely focused

Erik Tjong Kim Sang
Alfa-informatica, University of Groningen, Groningen, The Netherlands, e-mail: erikt@xs4all.nl

Katja Hofmann
ISLA, University of Amsterdam, Amsterdam, The Netherlands, e-mail: k.hofmann@uva.nl

Maarten de Rijke
ISLA, University of Amsterdam, Amsterdam, The Netherlands, e-mail: derijke@uva.nl

* Parts of this chapter have been published as (Tjong Kim Sang and Hofmann, 2007; Hofmann and Tjong Kim Sang, 2007; Tjong Kim Sang, 2009).

on English-language resources. Although WordNets exist for many languages, these are usually much smaller than Princeton WordNet (PWN) (Fellbaum, 1998), the major semantic resource for English.

For English, an excellent method has been developed for automatically extending lexical taxonomies. Snow et al (2005) show that it is possible to predict new and precise hypernym-hyponym relations from a parsed corpus. In this study, their approach is applied to data in another language (Dutch) and it is compared with alternative ways for deriving hypernym-hyponym relations. The following three research questions are examined:

1. How does the approach of Snow et al perform in comparison with single-pattern extraction methods applied to data resources of various sizes?
2. Snow et al relied on sense-tagged data which is presently unavailable for Dutch. What is the effect of this omission and how can one deal with it?
3. Snow et al relied on the availability of a full parser to preprocess their data, a tool which might be missing, take a lot of time to run or produce noisy results for other languages. What results can one obtain for the relation extraction task with available cheap and reliable shallow parser preprocessing?

After this introductory section, the task, the data and the evaluation metrics in the second section will be described. The next three sections present the findings concerning the three research questions presented above, comparison with the single-pattern extraction (Section 3), dealing with a missing word sense tagger (Section 4) and preprocessing with a shallow parser (Section 5). Section 6 is the conclusion.

2 Task and Approach

Techniques for automatically extending lexical resources such as WordNets are examined. This section describes which lexical relations are focused on, explains the necessary data preprocessing steps, describes the information being looked for and introduces the evaluation approach.

2.1 Task

The focus will be on extracting a particular lexical relation from text: hypernymy. One term is a hypernym of another if its meaning covers both an aspect of the meaning of the second term and is broader. For example, *furniture* is a hypernym of *table*. The opposite term for hypernym is hyponym. So *table* is a hyponym of *furniture*. Hypernymy is a transitive relation. If term A is a hypernym of term B while term B is a hypernym of term C then term A is also a hypernym of term C. Classes are also considered as hypernyms of their instances.

In Princeton WordNet (Fellbaum, 1998), hypernym relations are defined between senses of words. The Dutch WordNet (DWN), which is a part of EuroWordNet (Vossen, 1998), contains 659,284 of such hypernym noun pairs of which 100,268 are defined as hypernymy links and 559,016 are inherited by transitivity[2]. More importantly as the resource contains hypernym information for 45,979 different nouns. A test with a recent Dutch newspaper text revealed that the Dutch WordNet only covered about two-thirds of the noun lemmas in the newspaper (among the missing words were *e-mail*, *euro* and *provider*). Proper names, like names for people, organisations and locations, pose an even larger problem: DWN only contains 1608 words that start with a capital letter. Improving the coverage of DWN and similar resources is an important task. However, performing such a task manually is expensive, so it is interesting to know to what extent this task can be performed automatically. In the chapter by van der Plas, Fahmi, and Tiedemann, *Automatic Extraction of Medical Term Variants from Mutilingual Parallel Translations* (this volume), the automatic acquisition of another WordNet relation, synonymy, is investigated.

2.2 Natural Language Processing

Snow et al (2005) processed their text data with natural language processing tools like a full parser. However, these tools might not be available for other languages, might be expensive and time-consuming to run or might produce unreliable output. In this study different preprocessing techniques for texts are compared and their effects on subsequent relation extraction performance is evaluated.

The initial experiments will use as little preprocessing as possible. However, completely skipping the preprocessing step is not feasible. In the first study the following preprocessing steps to the source texts are applied:

- Tokenising: separating punctuation marks from words and identifying sentence boundaries;
- POS-tagging: assigning word classes to tokens;
- Lemmatising: assigning lemmas to tokens.

All three methods are cheap in processing time. Tokenising and lemmatising are useful because they map related different tokens to a single token. For example, tokenising would map *finishing.* to *finishing* and ., while lemmatising would map *finishing* to *finish*. This allows extraction patterns to be more general: only one pattern for *finish* is needed, rather than two extra patterns involving *finishing* and *finishing.*

For the patterns, it is very useful to be able to make a distinction between words of a different syntactic class, for example the noun and verb form of *finish*. Syntactic

[2] After the conclusion of the study, the Cornetto database (Vossen et al, 2008) has superseded Dutch Wordnet with among others a superior coverage of the Dutch lexicon.

classes are obtained by the POS-tagging step. It allows us to create extraction patterns which focus only on the noun occurrences of the target words, ignoring possible verb and adjective forms.

In the first study on the effects of pattern numbers and training data size, these three preprocessing techniques are employed only. In the second and third study full parsing is used. Apart from the three preprocessing steps mentioned earlier, this involves a fourth step in which the syntactic relation between pairs of words are computed. Related words might be neighbours in a sentence but they might also be words separated by several other words. This makes the full parsing step expensive in terms of computing power and computing time. This step has not been performed by ourselves but a text corpus has been relied on that was processed by the Dutch parser Alpino and was produced by the University of Groningen (van Noord, 2009).

In the first study, data is also retrieved with web queries which require plural forms of nouns. These forms are obtained from the plural list from CELEX (Baayen et al, 1995) (64,040 nouns). Words that are not present in the database receive a plural form, which is determined by a machine learner trained on the database. It has the seven final characters of the words as features, and can predict 152 different plural forms. Its leave-one-out accuracy on the training set is 89%.

2.3 Collecting Evidence

This study employed five strategies for finding evidence for hypernymy relations between pairs of words:

1. Assume that the longest known character suffix of the hyponym is a hypernym. This morphological approach maps *blackbird* to *bird* (Sabou et al, 2005). The suffix is required to be a known word and the prefix to contain at least three character;
2. Search a text for fixed text patterns like *A such as B and C* and consider these as evidence for relations between the word pairs *A* and *B*, and *A* and *C* (Hearst, 1992);
3. Search a text for fixed text patterns like *A, B and C* and consider these as evidence *A*, *B* and *C* having the same hypernym (Caraballo, 1999);
4. Use a combination of the results of strategies 2 and 3;
5. Starting from a list of seed pairs of related words like *A* - *B* search in the text for phrases between these words, for example *A xxx yyy B* and jointly use all of these phrases as patterns (like in 2) for finding other candidate pairs in the same context (*C xxx yyy D*) while assigning weights to the patterns based on how well they perform on the seed list (Snow et al, 2005).

The format of the text patterns that are being looked for depends on the method of text processing used. Two types of patterns are distingushed: dependency patterns, which are used in combination with preprocessing by full parsing, and lexical patterns, which are used in combination with shallow preprocessing.

Dependency patterns use the dependency relations between words that are generated by the Dutch Alpino parser. For example, the parser would produce the following relations between word pairs for the sentence *Large cities in northern England such as Liverpool are beyond revival*:

large:JJ:MOD:NN:*city*
city:NN:SUBJ:VBD:*be*
in:IN:MOD:NN:*city*
north:JJ:MOD:NNP:*England*
England:NNP:OBJ1:IN:*in*
such:DT:MOD:IN:*as*
as:IN:MOD:NN:*city*
Liverpool:NNP:OBJ1:IN:*as*
be:VB:–:–:–
beyond:IN:MOD:VB:*be*
revival:NN::OBJ1:IN:*beyond*

This analysis indicates that the word *large* is an adjective (JJ), which is a modifier (MOD) of head word *city*, that is a noun (NN). Punctuation has been separated from words and the relations contain lemmas rather than words. Based on this syntactic analysis, it is posisble to define dependency patterns like Snow et al (2005). Patterns are dependency paths between two nouns in the sentence with at most three intermediate nodes. Additional satellite nodes can be present next to the two nouns. Here is one of the patterns that can be derived for the two noun phrases *large cities* and *northern England* in the example sentence:

NP_1:NN:SUBJ:VBD:
in:IN:MOD:NN:NP_1
NP_2:NNP:OBJ1:IN:*in*

The pattern defines a path from the head lemma *city* via *in*, to *England*. Note that lexical information linking outside this pattern (*be* at the end of the first line) has been removed and that lexical information from the target noun phrases has been replaced by the name of the noun phrase (NP_1 on the first and second line). For each dependency pattern, six variants are built, four with additional information from the two noun phrases and two more with head information of one of the two target NPs. The pattern variants are similar to the lexical pattern variants that will be presented next.

The lexical patterns only use material generated by the tokenising, POS-tagging and lemmatising preprocessing steps. These would convert the example sentence to a sequence of lemmas with attached word class information: *large*/JJ *city*/NN *in*/IN *north*/JJ *England*/NNP *such*/DT *as*/IN *Liverpool*/NNP *be*/VB *beyond*/IN *revival*/NN ./. Again, *large* is an adjective (JJ) and *city* a noun (NN) but unlike with preprocessing by full parsing, the relation between the two words is undefined.

Lexical patterns contain two target phrases, both noun phrases. A maximum of three tokens can separate the two words. Additionally, the pattern may contain up to two optional extra tokens (a non-head token of the first noun phrase and/or one of the second noun phrase). The lexical preprocessing method uses two basic regular expressions for identifying noun phrases: *Determiner? Adjective* Noun+*

and *ProperNoun+*. It assumes that the final token of the matched phrase is the head. Here is one set of four patterns which can be derived from the example sentence, all related to the phrase *large cities in northern England*:

1. *NP in NP*
2. *large NP in NP*
3. *NP in north NP*
4. *large NP in north NP*

The patterns contain lemmas rather than the words of the sentence in order to allow for general patterns. For the same reason, the noun phrases have been replaced by the token *NP*. Each of the four patterns will be used as evidence for a possible hypernymy relation between the two noun phrase heads *city* and *England*. As a novel extension to the work of Snow et al., two additional variants of each pattern were included in which either the first NP or the second NP was replaced by its head:

5. *city in NP*
6. *NP in England*

Among others, this enabled us to identify patterns containing appositions: *president NP*.

When applying extraction strategies 2 and 3, the candidate hypernym most frequently occurring in a pattern with the source hyponym is selected. For strategy 4, the two frequencies in a hypernym evidence score $s(h,w)$, for each candidate hypernym h for word w are combined. This is the sum of the normalised evidence for the hypernymy relation between h and w, and the evidence for sibling relations between w and known hyponyms c of h:

$$s(h,w) = \frac{f_{hw}}{\sum_x f_{xw}} + \sum_c \frac{g_{cw}}{\sum_y g_{yw}}$$

where f_{hw} is the frequency of patterns that predict that h is a hypernym of w (strategy 2), g_{cw} is the frequency of patterns that predict that c is a sibling of w (strategy 3), and x and y are words from the wordnet. Each word w selects the candidate hypernym h with the largest score $s(h,w)$. Evidence for hypernyms and siblings is included in this score. Different scoring schemes are experimented with, for example, by including evidence from hypernyms of hypernyms (grandparents) and remote siblings (cousins), but this basic scoring scheme was found to perform best.

2.4 Evaluation

The hypernym extraction methods of this study are evaluated using the Dutch part of EuroWordNet (DWN) (Vossen, 1998). Hypernym-hyponym pairs that are present in the lexicon are assumed to be correct. In order for the evaluation to be complete, negative examples are also needed, pairs of words that are not related by hypernymy.

However, where DWN does not mention two words as being related by hypernymy, this could have several reasons. The words themselves might be missing from the resource, or their relation might not have been included.

In order to have negative examples for the evaluation, the same assumption as Snow et al (2005) is used: the hypernymy relation in DWN is complete for the words that it contains. This means that when two words are present in the lexicon without the target relation being specified between them, the assumption is made that they are unrelated. The presence of positive and negative relations allows for an automatic evaluation in which precision, recall and F values are computed.

The extraction method does not require finding the hypernym parent of a target word in DWN. Instead, the extraction approach returning any hypernym ancestor is enough. The two methods are used in order to rule out identification methods that simply return the top node of the hierarchy for all words. In study 1, the distance is being measured between the assigned hypernym and the target word. The ideal distance is one that would occur if the ancestor is a parent. A grandparent receives distance two and so on. And in study 3, the top node is simply disallowed for the hypernymy hierarchy as a candidate hypernym.

3 Study 1: Comparing Pattern Numbers and Corpus Sizes

This first study applies different hypernymy extraction methods to data sources of various sizes. First, a method for automatically deriving corpus-specific extraction patterns is evaluated from a set of examples. Subsequently, a method for combining these patterns is examined, and a comparison is made between the performance of the combination with the best individual patterns and the morphological approach described in Section 2.4. Finally, two fixed patterns are applied to web data, and their performance compared with the earlier corpus results. The study concludes with an analysis of the errors made by the best system .

3.1 Extracting Individual Patterns

In this study, the Twente Nieuws Corpus was used, a corpus of Dutch newspaper text and subtitle text covering four years (1999–2002) and containing about 300M words. The corpus was processed by automatic tools which tokenised it, assigned POS tags and identified lemmas. Next the same approach as Snow et al (2005) is used but with lexical information rather than dependency parses: all pairs of nouns with four or fewer tokens (words or punctuation signs) between them were selected. The intermediate tokens (labelled `infix`) as well as the token before the first noun (`prefix`) and the token following the second noun (`suffix`) were stored as a pattern. For each noun pair, four patterns were identified:

- `N1 infix N2`

- `prefix N1 infix N2`
- `prefix N1 infix N2 suffix`
- `N1 infix N2 suffix`

Table 1 Top ten high precision patterns of the format `N1 infix N2` extracted from the text corpus which have a recall score higher than 0.00100. In the patterns, N-pl and N-sg represent a plural noun and a singular noun, respectively. It is possible to aggregate patterns by ignoring the number of the noun (N-pl + N-sg = N) in order to achieve higher recall scores at the expense of lower precision rates. The phrase in parentheses is an English translation of the main words of the pattern.

Precision	Recall	$F_{\beta=1}$	Dist,	Pattern
0.375	0.00137	0.00273	2.56	N-pl , vooral N-pl (*especially*)
0.300	0.00133	0.00264	2.23	N-pl , waaronder N-pl (*among which*)
0.258	0.00120	0.00238	1.55	N-pl , waaronder N-sg (*among which*)
0.250	0.00196	0.00388	2.08	N-pl of ander N-pl (*or other*)
0.244	0.00418	0.00821	1.96	N-pl zoals N-sg (*such as*)
0.220	0.00259	0.00512	2.10	N-pl zoals N-pl (*such as*)
0.213	0.00809	0.01559	1.99	N-pl en ander N-pl (*and other*)
0.205	0.00387	0.00760	2.20	N-pl , zoals N-pl (*such as*)
0.184	0.00396	0.00775	1.78	N-pl , zoals N-sg (*such as*)
0.158	0.00394	0.00768	1.68	N-sg en ander N-pl (*and other*)

The patterns also included information about whether the nouns were singular or plural, a feature which can be derived from the POS tags. 3,283,492 unique patterns were identified. The patterns were evaluated by registering how often they assigned correct hypernym relations corresponding to noun pairs from DWN. Only 118,306 patterns had a recall that was larger than zero. The majority of these patterns (63%) had a precision of 1.0 but the recall of these patterns was very low (0.00003-0.00025). The highest registered recall value for a single pattern was 0.00897 (for *N-pl and N-pl*). The recall values are low because of the difficulty of the task: it was aimed at generating a valid hypernym for *all* 45,979 nouns in the Dutch WordNet. A recall value of 1.0 corresponds with single pattern predicting a correct hypernym for every noun in DWN, something which is impossible to achieve.

Table 1 lists ten top-precision patterns of the format `N1 infix N2` and a precision score of 0.158 or higher. Figure 1 contains an overview of the precision and recall values of all 421 patterns of that group. For comparison with other approaches, the pattern *N zoals N* has been selected, a combination of the results of four patterns of which two are listed in Table 1. This pattern obtained a precision score of 0.22 and a recall score of 0.0068 (Table 2).

3.2 Combining Corpus Patterns

Snow et al (2005) showed that for the task of collecting hypernym-hyponym pairs, a combination of extraction patterns outperform the best individual pattern. In order to obtain a combined prediction of a set of patterns, they represented word pairs

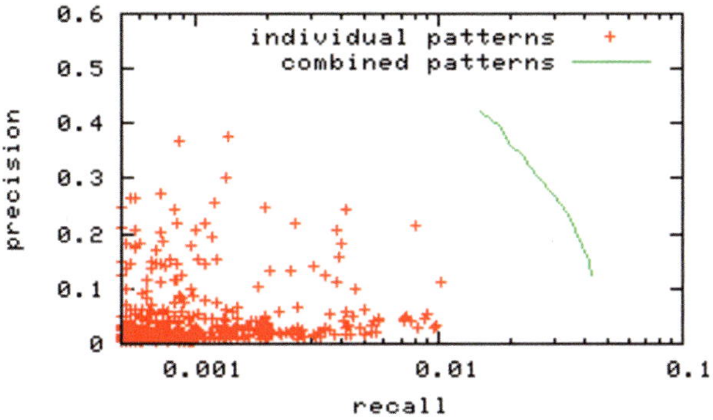

Fig. 1 Precision and recall values of the 421 hypernym-hyponym extraction patterns of the format `N1 infix N2` with the highest recall values when applied to the text corpus (+) compared with combinations of these patterns (line). Pattern combinations outperform individual patterns both with respect to precision and recall. The recall values are low because of the difficulty of the task (reproducing valid hypernyms for all nouns in the wordnet being used).

Table 2 Performance measured with the corpus approach and the morphological approach. The pattern combination perform better than the best individual pattern but both suffer from low recall figures. The conjunctive pattern and the morphological approach, predicting the longest known suffix of each word as its hypernym (Section 2.4), surprisingly enough outperform both corpus approaches on most evaluation measures.

Method	**Prec.**	**Recall**	**$F_{\beta=1}$**	**Dist.**
Corpus: *N zoals N*	0.22	0.0068	0.013	2.01
Corpus: combined	0.36	0.020	0.038	2.86
Corpus: *N en N*	0.31	0.14	0.19	1.98
Morphological approach	0.54	0.33	0.41	1.19

by a sequence of numeric features. The value of each feature was determined by a single pattern predicting that the word pair was related according to the hypernymy relation or not. A machine learning method, Bayesian Logistic Regression was used to determine the combined prediction of feature sets for unknown word pairs based on a comparison with known word pairs which could be part of the relation or not.

This work of Snow et al (2005) for the Dutch data has been replicated. 16728 features have been identified which correspond with hypernym-hyponym extraction patterns. All noun pairs which were associated with at least five of these patterns in the text corpus, were represented by numerical features which encoded the fact that the corresponding pattern predicted that the two were related (value 1) or not (value 0). Only nouns present in the Dutch WordNet (DWN) were considered. The class associated with each feature set could either be positive if the ordered word pair occurred in the hypernymy relation of DWN or negative if the ordered pair was not in the DWN relation. This resulted in a dataset of 528,232 different ordered pairs of which 10,653 (2.0%) were related.

The performance of the combined patterns was determined by 10-fold cross validation: the training set was divided into ten parts and the classes for each part were predicted by using the other nine parts as training data. Like Snow et al (2005) Bayesian Logistic Regression was used as a learning technique (Genkin et al, 2004). Support Vector Machines have also been tested but these proved to be unable to process the data within a reasonable time.

The classifier assigned a confidence score between 0 and 1 to each pair. The precision and recall values were computed for different acceptance threshold values (0.001-0.90) which resulted in the line in Figure 1. The combined patterns obtain similar precision scores as the best individual patterns but their recall scores are a lot higher. For comparison with other approaches, the acceptance threshold 0.5 has been used, which resulted in a precision of 0.36 and a recall of 0.020 (Table 2).

Contrary to the expectation, both alternative hypernym prediction methods outperform the combination of lexical patterns (Table 2). The conjunctive pattern obtains a lower precision score than the combination but its recall is an order of larger magnitude. The morphological approach of selecting the shortest suffix that is also a valid word as the candidate hypernym (*blackbird* → *bird*), does even better: obtaining precision, recall and distance scores that are the best of all examined approaches. Although the morphological approach is useful for deriving relations between morphological variants, it cannot be applied for finding every hypernym-hyponym pair. For example, it cannot find out that a *poodle* is a *dog* because the latter word is not part of the former. The relation between word pairs like these need to be identified by another approach.

3.3 Web Query Format

Next, the web will be searched for lexical patterns as evidence of lexical relations. When working with a fixed corpus on disk, an exhaustive search can be performed. For web search, however, this is not possible. Instead, acquiring interesting lexical patterns from text snippets returned for specific queries is relied on. The format of the queries has been based on three considerations.

First, a general query like * *such as* * is insufficient for obtaining much interesting information. Today, most web search engines impose a limit on the number of results returned from a query (for example 1000), which limits the opportunities for assessing the performance of such a general pattern. In order to obtain useful information, the query needs to be more specific. For the pattern * *such as* *, two options are possible: adding the hypernym, which gives *hypernym such as* *, or adding the hyponym, which results in * *such as hyponym*.

Both extensions of the general pattern have their disadvantages. A pattern that includes the hypernym may fail to generate much useful information if the hypernym has many hyponyms. And patterns with hyponyms require more queries than patterns with hypernyms (at least one per child rather than one per parent). Hyponyms are included in the patterns. This approach models the real-world task in which someone is looking for the meaning of an unknown entity.

The final consideration concerns the hyponyms to be used in the queries. The focus is on evaluating the approach via comparison with an existing wordnet. Rather than flooding the search engine with queries representing every hyponym in the lexical resource, only a random sample of hypernyms was chosen to be searched for. The evaluation score was observed to converge for approximately 1500 words and this is the number of queries settled for.

3.4 Web Extraction Results

For this web extraction work, two fixed context patterns are used: one containing the word *zoals* (*such as*), a reliable and reasonably frequent hypernym pattern according to the corpus work, and another containing the word *en* (*and*), the most frequent pattern found in the text corpus. Candidate hyponyms are chosen to be randomly added to the queries to improve the chance of retrieving interesting information.

This approach worked well. As Table 3 shows, both patterns outperformed the F-score of the combined patterns in the corpus experiments. As in the corpus experiments, the conjunctive web pattern outperformed the *such as* web pattern with respect to precision and recall. It is assumed that the frequency of the two patterns plays an important role (the Google index contains about five times as many pages with the conjunctive pattern as *zoals*).

Finally, word-internal information was combined with the conjunctive pattern approach, by adding the morphological candidates to the web evidence before computing hypernym pair scores. This approach achieved the highest recall, with only a slight loss in precision (Table 3). A basic combination approach of using the conjunctive pattern for searching for hyponyms for which no candidates were generated by the morphological approach, would have achieved a similar performance.

Table 3 Performance measured in the two web experiments and a combination of the best web approach with the morphological approach. The conjunctive web pattern *N en N* rates best, because of its high frequency. All evaluation scores can be improved by supplying the best web approach with word-internal information.

Method	**Prec.**	**Recall**	$\mathbf{F}_{\beta=1}$	**Dist.**
Web: *N zoals N*	0.23	0.089	0.13	2.06
Web: *N en N*	0.39	0.31	0.35	2.04
Morphological approach	0.54	0.33	0.41	1.19
Web: *en* + morphology	0.48	0.45	0.46	1.64

3.5 Error Analysis

The output of the conjunctive web extraction with word-internal information was inspected. For this purpose the ten most frequent hypernym pairs (top group, see Table 4) were selected, the ten least frequent (bottom group) and the ten pairs exactly between these two groups (centre group): 40% of the pairs were correct, 47% incorrect and 13% were plausible but contained relations that were not present in the reference wordnet. In the centre group all errors were caused by the morphological approach while all other errors in the top group and in the bottom group originated from the web extraction method.

3.6 Discussion

The contributions of the first study are two-fold. First, it showed that the large quantity of available web data allows basic patterns to perform better on hypernym extraction than an advanced combination of extraction patterns applied to a large corpus. Second, it is demonstrated that the performance of the web extraction method can be improved by combining its results with those of a corpus-independent morphological approach.

While the web results are of reasonable quality, some concern can be expressed about the quality of the corpus results. At best, an F-value of 0.038 is obtained, a lot lower than the 0.348 reported for English in (Snow et al, 2005). There are two reasons for this difference. First, the evaluation methods are different: the aim is to generate hypernyms for all words in the wordnet that are used while Snow et al. only look for hypernyms for words in the wordnet *that are present in their corpus*. Second, in their extraction work Snow et al. also use a sense-tagged corpus, a resource which is unavailable for Dutch at present. In the next section, this difference will be examined in more detail.

Table 4 Example output of the conjunctive web system with word-internal information. Of the ten most frequent pairs, four are correct (+). Four others are plausible but are missing (?) in the wordnet used for evaluation.

+/-	score	hyponym	hypernym
-	912	buffel	predator
+	762	trui	kledingstuk
?	715	motorfiets	motorrijtuig
+	697	kruidnagel	specerij
-	680	concours	samenzijn
+	676	koopwoning	woongelegenheid
+	672	inspecteur	opziener
?	660	roller	werktuig
?	654	rente	verdiensten
?	650	cluster	afd.

4 Study 2: Examining the Effect of Ambiguity

In the second study, the relation extraction method of Snow et al (2005) is applied to a non-English language (Dutch) for which only a basic wordnet and a parser are available. Without an additional sense-tagged corpus it is found that this approach is highly susceptible to noise due to word sense ambiguity. Two methods to address this problem are proposed and evaluated.

4.1 Approach

This approach closely follows the one described in Snow et al (2005). From a parsed newspaper corpus, all sentences are selected which contain nouns found in the Dutch part of EuroWordNet (DWN) (Vossen, 1998). Ordered noun pairs from each sentence are extracted and the noun pair labelled as *Known Hypernym*, if they are related according to the transitive closure of the hypernym relation in DWN. All other noun pairs occurring in DWN are labelled *Known Non-Hypernym* and conform to the completeness assumption described in Section 2.4. Next, all dependency paths are collected between noun pairs and used as features in a machine learning experiment for predicting Known Hypernym pairs. The paths that occur with fewer than five unique noun pairs are ignored, as well as noun pairs which appear with fewer than five different dependency paths.

As seen in the previous section, the initial implementation performed much worse than the results reported in Snow et al (2005) (F-score of 0.11 vs. 0.348). A major difference between the two systems, was that Snow et al (2005) use frequency counts for word-senses from a sense-tagged corpus to reduce ambiguity in their training data. As no such resource is available for Dutch, alternative ways of reducing ambiguity in the training data were explored. Two methods for reducing ambiguity are proposed and performance at different levels of ambiguity is compared. First, assume that DWN assigns the first sense (i.e., lowest sense number) to the most frequent sense of a polysemous word. Following this assumption, filter training instances labelled as *Known Hypernym* to only include noun pairs where a hypernym relation exists for the first senses of both nouns. The second method circumvents the problem of ambiguity, by including only those training instances where both nouns are monosemous (i.e., have only one sense). This restriction avoids regarding contexts as interesting, where the target word pairs are only related via other senses.

4.2 Experiments and Results

These experiments are based on a part of the Twente News Corpus: (TwNC-02), consisting of over 23 million sentences from Dutch newspaper articles. The corpus was parsed with Alpino (van der Beek et al, 2002). Dependency patterns are generated from the parses by moving the category of the *head* child of each node into the parent node. Satellite links are added to the words preceding and following the pattern in the sentence. Thus, each word pair extracts up to four patterns.

Using all DWN nouns, the ratio of negative to positive examples in the corpus is approximately 38:1 compared to 50:1 in Snow et al (2005). The relatively high number of positive instances are attributed to a large number of false positives due to word-sense ambiguity. For the reduced-ambiguity data sets the percentage of negative instances is much higher. Like Snow et al (2005), the ratio of negative to positive items is restricted in these data sets to 50:1 by randomly removing negative items.

A Logistic Regression[3] classifier is trained to predict hypernym relations between word pairs using binary features. Each feature represents the occurrence of a dependency path for a given noun pair. A single data set which is evaluated using 10-fold cross validation is worked with.

Five experiments were performed. In (a) all 45,981 nouns from DWN were included. (b) used only positive instances where the hypernym relation was applied to the first sense of each noun. (c) was restricted to the 40,069 monosemous DWN nouns. In order to cancel the benefits of larger training sets, the size of the first three data sets was limited to the size of the smallest, (c), through random selection of training instances. Experiments (d) and (e) were performed with the complete data sets of (a) and (b) respectively.

Table 5 shows precision, recall and $F_{\beta=1}$ of the machine learner for each data set. Results obtained with the monosemous data set are significantly ($p \ll 0.001$, estimated with bootstrap resampling) better than those obtained with all data, even for data sets which are considerably larger. Figure 2 shows that the largest number of high-precision patterns was identified using the monosemous data set.

Table 5 Results from training with: (a+d) all DWN noun data; (b+e) only the first sense of each noun; and (c) monosemous nouns only.

	# pairs	# pos	# neg	Prec.	Rec.	$F_{\beta=1}$
a	253,068	5,108	247,960	0.067	0.078	0.072
b	250,786	5,041	245,745	0.148	0.154	0.151
c	251,127	5,115	246,012	**0.215**	**0.302**	**0.251**
d	1,718,369	43,267	1,675,102	0.085	0.160	0.111
e	1,093,150	21,035	1,072,115	0.136	0.209	0.165

The best F-score (0.251) is lower than the best score reported in (Snow et al, 2005) (0.348). It is suspected that the prime reason for this is the size difference between the bootstrap lexical resources: PWN (114,648 nouns) and DWN (45,981). This allows the English work to derive many more positive sense-restricted examples (14,387 in comparison with the 5,115 for Dutch) even though access is given to a larger parsed corpus (23 million vs. 6 million sentences).

[3] `http://stat.rutgers.edu/$\sim$madigan/BBR/`

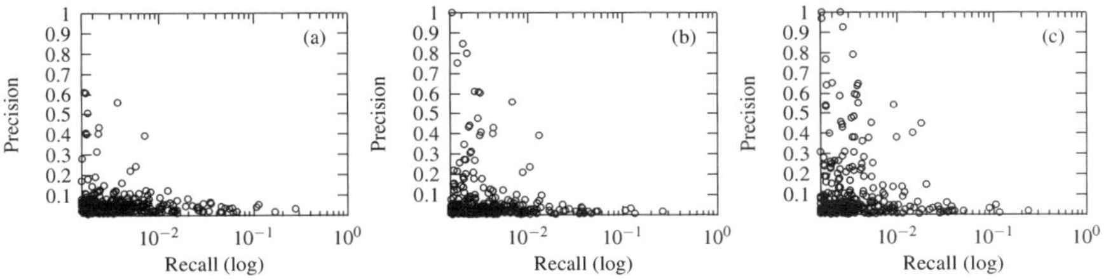

Fig. 2 Predictive quality of individual dependency patterns extracted using (a) all nouns in DWN; (b) first noun senses only; and (c) monosemous words only. Each data point represents precision and recall of one dependency pattern.

4.3 Discussion

The contributions of this second study are two-fold. First, it confirms the results of Snow et al (2005) and shows that the method of automatically extracting dependency patterns for hypernym acquisition can be transferred to languages other than English. Second, that the method is sensitive to word sense ambiguity is shown but that this problem can be addressed by removing polysemous nouns from the training data.

The results obtained in this study improve on earlier work on extracting Dutch hypernym relations. IJzereef (2004) reports a precision score of 0.198 for extracting hypernym relations with a small set of fixed patterns (Table 5.3, recall figures have not been included). van der Plas and Bouma (2005) use a more complex evaluation score which cannot easily be compared with the evaluation approach.

Considerable follow up work could be carried out, ranging from further evaluation, to applications of the method in broader contexts. The next step for this team will be manual evaluation of the method, as a comparison with DWN is not sufficient for assessing its actual performance. Further evaluation will also allow the assessment of the sensitivity of the approach to factors other than ambiguity, for example corpus size, differences in dependency parses, or the use of lexical patterns instead of dependency parses.

Another promising direction for future work would be to exploit features specific to the Alpino dependency parser, such as multiword phrases. These elements usually contain named entities, which would be interesting to add to DWN as it currently contains few named entities. This method of automatic dependency pattern extraction is planned to apply to extending DWN and by using it for deriving relations other than hypernymy.

An improvement of the results is expected from using the new Cornetto database (Vossen et al, 2008) as a lexical source rather than the Dutch WordNet. Cornetto's size, which is expected to be similar to PWN, could prove to be an important factor for decreasing the performance differences with the hypernymy extraction studies applied to English.

5 Study 3: Measuring the Effect of Syntactic Processing

In the third and final study, the effect of two text preprocessing approaches on the task of extracting hypernymy information is examined. The first of the two methods is shallow linguistic processing, a robust and fast text analysis method which only uses information from words, like lemma information and POS classes. The second method is dependency parsing, which includes information about the syntactic relations between words. The task, the preprocessing methods and the evaluation setting have already been described in Sections 2.1, 2.2 and 2.4, respectively. In the next section, how the experiments were set up and the results were presented. After that the effect of the two methods on the extraction task will be examined.

5.1 Experiments and Results

The extraction techniques to two different Dutch corpora have been applied. The first is a collection of texts from the news domain. It consists of texts from five different Dutch newspapers from the Twente News Corpus collection. Two versions of this corpus exist. The version which contains the years 1997-2005 (26 million sentences and 450 million tokens) have been worked with. The second corpus is the Dutch Wikipedia. Here a version of October 2006 (5 million sentences and 58 million words) is used.

Syntactic preprocessing of the material was done with the Alpino parser, the best available parser for Dutch with a labelled dependency accuracy of 89% (van Noord, 2006). Rather than performing the parsing task ourselves, an available parsed treebank was relied on which included the text corpora that was to be used. (van Noord, 2009).

The parser also performs POS-tagging and lemmatisation tasks that are useful for the lexical preprocessing methods. However, keeping future real-time applications into account, the lexical processing did not want to be dependent on the parser. Therefore an in-house POS tagger was developed and a lemmatiser based on the material created in the Corpus Spoken Dutch project (Van Eynde, 2005). The tagger achieved an accuracy of 96% on test data from the same project while the lemmatiser achieved 98%.

The Dutch part of EuroWordNet (Vossen, 1998) was used as the gold standard lexical resource, both for training and testing. In the lexicon, many nouns have different senses. This can cause problems for the pattern extraction process. For example, if a noun N_1 with sense X is related to another noun N_2 then the appearance of N_1 with sense Y with N_2 in the text may be completely accidental and say nothing about the relation between the two words. In that case it would be wrong to regard the context of the two words as an interesting extraction pattern. This problem was avoided by using the approach described in the previous section, removing all nouns with multiple senses from the data set and using only the monosemous words for finding good extraction patterns. This restriction is only imposed in the training phase. Both monosemous words and polysemous words are considered in the evaluation process.

Two additional restrictions are imposed on the lexical resource. First, the top noun of the hypernymy hierarchy (*iets*) was removed from the list of valid hypernyms. This word is a valid hypernym of any other noun. It is not an interesting suggestion for the extraction procedure to put forward. Second, the extraction procedure was restricted to propose only known hypernyms as candidate hypernyms. Nouns that appeared in the lexical resources only as hyponyms (leaf nodes of the hypernymy tree) were never proposed as candidate hypernyms. This made sense for the evaluation procedure which is only aimed at finding known hypernym-hyponym pairs.

Table 6 Hypernym extraction scores for the five newspapers in the Twente News Corpus (AD, NRC, Parool, Trouw and Volkskrant) and for the Dutch Wikipedia. The Targets column shows the number of unique positive word pairs in each data set. The Dutch Wikipedia contains about as much data as one of the newspaper sections.

Lexical patterns

Data source	Targ.	Prec.	Recall	$F_{\beta=1}$
AD	620	55.8%	27.9%	37.2
NRC	882	50.4%	23.8%	32.3
Parool	462	51.8%	21.9%	30.8
Trouw	607	54.1%	25.9%	35.0
Volkskrant	970	49.7%	24.1%	32.5
Newspapers	3307	43.1%	26.7%	33.0
Wikipedia	1288	63.4%	44.3%	52.1

Dependency patterns

Data source	Targ.	Prec.	Recall	$F_{\beta=1}$
AD	706	42.9%	30.2%	35.4
NRC	1224	26.2%	25.3%	25.7
Parool	584	31.2%	23.8%	27.0
Trouw	760	35.3%	29.0%	31.8
Volkskrant	1204	29.2%	25.5%	27.2
Newspapers	3806	20.7%	29.1%	24.2
Wikipedia	1580	61.9%	47.0%	53.4

Two hypernym extraction experiments were performed, one which used lexical extraction patterns, and one which used dependency patterns.[4] The results from the experiments can be found in Table 6. The newspaper F-scores obtained with lexical patterns are similar to those reported for English (Snow et al, 2005, 32.0) but the dependency patterns perform worse. Both approaches perform well on Wikipedia data, most likely because of the more repeated sentence structures and the presence of many definition sentences. For newspaper data, lexical patterns outperform dependency patterns, both for precision and $F_{\beta=1}$. For Wikipedia data, the differences are smaller and in fact the dependency patterns obtain the best F-score. For all data sets, the dependency patterns suggest more related pairs than the lexical patterns (column Targets). The differences between the two pattern types are significant ($p < 0.05$) for all evaluation measures for newspapers and for positive targets and recall for Wikipedia.

[4] The software used in these experiments has been made available at `http://www.let.rug.nl/erikt/cornetto/D08.zip`.

5.2 Result Analysis

This section takes a closer look at the results described in the previous section. An explanation for the differences between the scores obtained with lexical patterns and dependency patterns is looked for at the start. First the results for Wikipedia data were examined and then the results for newspaper data. Finally, an error analysis is performed to find out the strengths and weaknesses of each of the two methods.

The most important difference between the two pattern types for Wikipedia data is the number of positive targets (Table 6). Dependency patterns find 23% more related pairs in the Wikipedia data than lexical patterns (1580 vs. 1288). This effect can also be simulated by changing the size of the corpus. If the data set of the dependency patterns is restricted to 70% of its current size then the patterns retrieve a similar number of positive targets as the lexical patterns, 1289, with comparable precision, recall and $F_{\beta=1}$ scores (62.5%, 46.6% and 53.4). So the effect of applying the dependency patterns is expected to be the same as applying the lexical patterns to 43% more data.

Performance-wise there seems to be only a small difference between the two preprocessing methods when applied to the Wikipedia data set. However, when the scores obtained on the newspaper data (Table 6) are examined, larger differences are found. Dependency patterns find more positive targets and obtain a larger recall score, but their precision score is disappointing. However, when the precision-recall plots of the two methods were examined (Figure 3, obtained by varying the acceptance threshold of the machine learner), they were almost indistinguishable. The performance line for lexical patterns extends further to the left than the one for the dependency patterns, however the remainder of the two graphs overlap. The measured performances in Table 6 are different because the machine learner put the acceptance level for extracted pairs at different points on the graph: the performance lines in both newspaper graphs contain (recall,precision) points (26.7%,43.1%) and (29.1%,20.7%).

Major differences in the results of the two approaches are unable to be found. It can be concluded that, apart from an effect which can be simulated with some extra data, there is no difference between preprocessing text with shallow methods and with a full dependency parser.

Despite the lack of performance differences between the two preprocessing methods, there are still internal differences which cause one method to generate different related word pairs than the other. Next, two extraction patterns will be examined in detail and specify their distinct effects on the output results. It is hoped that by carefully examining their output the strengths and weaknesses of the two approaches can be learned.

A closer look at extraction pattern *N such as N* is taken for Newspaper data (second best for lexical patterns and fifth best for dependency patterns, see Table 7). The lexical pattern found 222 related word pairs while the dependency pattern discovered 199. 118 of these pairs were found by both patterns which means that the lexical pattern missed 81 of the pairs while the dependency pattern missed 104.

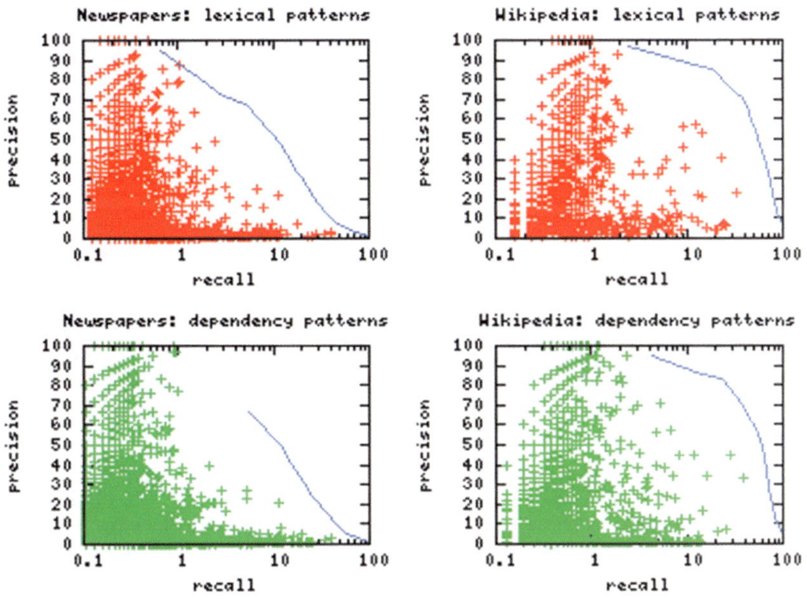

Fig. 3 Performance of individual hypernym extraction patterns applied to the combination of five newspapers and Wikipedia. Each + in the graphs representing a different extraction pattern. The precision-recall graphs for the machine learner (lines) are identical for each data source except for the extended part of the performance line for lexical patterns.

Table 7 Best performing extraction patterns according to F-scores.

lexical patterns applied to Newspapers

Key Phrase	Targ.	Prec.	Recall	$F_{\beta=1}$
N and other N	376	22.0%	11.4%	15.0
N such as N	222	25.1%	6.7%	10.6
N like N	579	7.6%	17.5%	10.6
N , such as N	263	15.6%	8.0%	10.5
N (N	323	7.5%	9.8%	8.5

Dependency patterns applied to Newspapers

Key Phrase	Targ.	Prec.	Recall	$F_{\beta=1}$
N and other N	420	21.1%	11.0%	14.5
N be a N	451	8.2%	11.8%	9.7
N like N	205	27.3%	5.4%	9.0
N be N	766	5.7%	20.1%	8.8
N such as N	199	22.4%	5.2%	8.5

Lexical patterns applied to Wikipedia

Key Phrase	Targ.	Prec.	Recall	$F_{\beta=1}$
N be a N	294	40.8%	22.8%	29.3
N be N	418	22.9%	32.5%	26.9
a N be N	185	53.3%	14.4%	22.6
N such as N	161	57.5%	12.5%	20.5
N (N	188	21.2%	14.6%	17.3

Dependency patterns applied to Wikipedia

Key Phrase	Targ.	Prec.	Recall	$F_{\beta=1}$
N be N	609	33.6%	38.5%	35.9
N be a N	452	44.3%	28.6%	34.8
the N be N	258	34.0%	16.3%	22.1
a N be N	184	44.7%	11.6%	18.5
N N	234	16.6%	14.8%	15.6

An overview of the cause of the recall errors can be found in Table 8. The two extraction patterns do not overlap completely. The dependency parser ignored punctuation signs and therefore the dependency pattern covers both phrases with and without punctuation. However, these phrase variants result in different lexical patterns. This is the cause for 56 hypernyms being missed by the lexical pattern. Meanwhile there is a difference between a dependency pattern without the conjunction *and* and one with the conjunction, while there is a unified lexical pattern processing both phrases with and without conjunctions. This caused the dependency pattern to miss 45 hypernyms. However, all of these 'missed' hypernyms are handled by other patterns.

The main cause of the recall differences between the two extraction patterns was the parser. The dependency pattern found twelve hypernyms which the lexical pattern missed because they required an analysis which was beyond POS-tagging and the basic noun phrase identifier used by the lexical preprocessor. Six hypernyms required extending a noun phrase with a prepositional phrase, five needed noun phrase extension with a relative clause and one involved appositions. An example of such a phrase is *illnesses caused by vitamin deficits, like scurvy and beriberi*.

Table 8 Primary causes of recall errors made by the lexical pattern *N such as N* (left) and the best performing corresponding dependency pattern (right).

56	—	covered by other patterns	45	—	covered by other patterns
12	48%	required full parsing	38	64%	parsing errors
6	24%	lemmatisation errors	10	17%	lemmatisation errors
3	12%	omitted for lack of support	7	12%	extraction pattern errors
3	12%	Pos-tagging errors	3	5%	omitted for lack of support
1	4%	extraction pattern error	1	2%	Pos-tagging error
81	100%		104	100%	

However, the syntactic information that was available to the dependency pattern did also have a negative effect on its recall: 38 of the hypernyms detected by the lexical pattern were missed by the dependency pattern because there was a parsing error in the relevant phrase. In more than half of the cases, this involved attaching the phrase starting with *such as* at an incorrect position. It was discovered that a phrase like N_1 *such as* N_2 , N_3 *and* N_4 could have been split in any position. Even some cases of prepositional phrases were found, and it was discovered that some relative clauses had been incorrectly moved from other positions in the sentence into the target phrase.

Other recall error causes appear less frequently. The two preprocessing methods used different lemmatisation algorithms which also made different errors. The effects of this were visible in the errors made by the two patterns. Some hypernyms were found by both patterns but were not present in both results because of insufficient support from other patterns (candidate hypernyms should be supported by at least five different patterns). The effect of errors in POS tags was small. The data analysis also revealed some inconsistencies in the extraction patterns, which should be examined.

5.3 *Discussion*

The effects of two different preprocessing methods for a natural language processing task has been evaluated: automatically identifying hypernymy information. The first method used lexical patterns and relied on shallow processing techniques like POS-tagging and lemmatisation. The second method used dependency patterns which relied on additional information obtained from dependency parsing.

In earlier work, McCarthy et al (2007) found that word sense disambiguation using thesauri generated from dependency relations performed only slightly better than thesauri generated from proximity-based relations. Jijkoun et al (2004) showed that information obtained from dependency patterns significantly improved the performance of a QA system. Li and Roth (2001) report that preprocessing by shallow parsing allows for a more accurate post-processing of ill-formed sentences than preprocessing with full parsing.

The study supports the findings of McCarthy et al (2007). Only minor differences in performances between the two preprocessing methods have been found. The most important difference: about 20% extra positive cases that were identified by the dependency patterns applied to Wikipedia data, can be overcome by increasing the data set of the lexical patterns by half. Obtaining more data may often be easier than dealing with the extra computing time required for parsing the data. For example, in the course of performing this study, a recent version of Wikipedia was not used because parsing the data would have taken 296 *days* on a single processor machine compared with a single hour for tagging the data.

6 Concluding Remarks

Three studies in automatic relation extraction were presented. The first study showed that fixed extraction patterns applied to web data were able to extract more related words than a set of derived patterns applied to corpus data. Rules based on the morphological structure of words perform even better, but can only be applied for word pairs which are morphologically related. The best-performing extraction approach was a combination of web patterns and morphological patterns.

In the second study, the effect of ambiguity on the extraction process was examined. A distinctive effect was observed: the results improved when positive examples related to non-prominent word senses were removed from the training data. Results got even better when all ambiguous words were removed from the training data altogether. These two approaches are very suitable for performing relation extraction for languages for which no word-sense tagger is available.

The third study measured the effect of including elaborate syntactic information in the extraction process. Only a small positive effect of this information was found, one which could also be obtained with a combination of shallow syntactic processing and a limited amount of extra training data. This is also a good observation for future application of these extraction techniques to languages for which only few language processing tools are available.

Acknowledgements

The research leading to these results was supported by the Netherlands Organisation for Scientific Research (NWO) under project nrs 264.70.050 (IMIX: FactMine), 612.066.512 (QASSIR), 612.061.814 (CLiKS), 612.061.815 (TNT) and 640.004.-802 (Bridge), by the Cornetto and DuOMAn projects carried out within the STEVIN programme which is funded by the Dutch and Flemish Governments (http://www.stevin-tst.org), and by the European Union's ICT Policy Support Programme as part of the Competitiveness and Innovation Framework Programme, CIP ICT-PSP under grant agreement nr 250430.

References

Baayen R, Piepenbrock R, Gulikers L (1995) The CELEX Lexical Database (Release 2) [CD-ROM]. Philadelphia, PA: Linguistic Data Consortium, University of Pennsylvania

van der Beek L, Bouma G, Malouf R, van Noord G (2002) The alpino dependency treebank. In: Proceedings of CLIN 2001, Twente University

Caraballo SA (1999) Automatic construction of a hypernym-labeled noun hierarchy from text. In: Proceedings of ACL-99, Maryland, USA

Fellbaum C (1998) WordNet – An Electronic Lexical Database. The MIT Press

Genkin A, Lewis DD, Madigan D (2004) Large-Scale Bayesian Logistic Regression for Text Categorization. Technical report, Rutgers University, New Jersey

Hearst MA (1992) Automatic acquisition of hyponyms from large text corpora. In: Proceedings of ACL-92, Newark, Delaware, USA

Hofmann K, Tjong Kim Sang E (2007) Automatic extension of non-english wordnets. In: Proceedings of SIGIR'07, Amsterdam, The Netherlands, (poster)

IJzereef L (2004) Automatische extractie van hyperniemrelaties uit grote tekstcorpora. MSc thesis, University of Groningen, (in Dutch)

Jijkoun V, de Rijke M, Mur J (2004) Information extraction for question answering: Improving recall through syntactic patterns. In: Proceedings of Coling'04, Geneva, Switzerland

Li X, Roth D (2001) Exploring evidence for shallow parsing. In: Proceedings of Conference on Computational Natural Language Learning (CoNLL) 2001

McCarthy D, Koeling R, Weeds J, Caroll J (2007) Unsupervised acquisition of predominant word senses. Computational Linguistics 33(4)

van Noord G (2006) At last parsing is now operational. In: Mertens P, Fairon C, Dister A, Watrin P (eds) TALN06. Verbum Ex Machina. Actes de la 13e conference sur le traitement automatique des langues naturelles

van Noord G (2009) Huge parsed corpora in lassy. In: Proceedings of TLT7, LOT, Groningen, The Netherlands

van der Plas L, Bouma G (2005) Automatic acquisition of lexico-semantic knowledge for qa. In: Proceedings of the IJCNLP Workshop on Ontologies and Lexical Resources, Jeju Island, Korea

Sabou M, Wroe C, Goble C, Mishne G (2005) Learning domain ontologies for web service descriptions: an experiment in bioinformatics. In: 14th International World Wide Web Conference (WWW2005), Chiba, Japan

Snow R, Jurafsky D, Ng AY (2005) Learning syntactic patterns for automatic hypernym discovery. In: NIPS 2005, Vancouver, Canada

Tjong Kim Sang E (2009) To use a treebank or not – which is better for hypernym extraction. In: Proceedings of the Seventh International Workshop on Treebanks and Linguistic Theories (TLT 7), Groningen, The Netherlands

Tjong Kim Sang E, Hofmann K (2007) Automatic extraction of dutch hypernym-hyponym pairs. In: Proceedings of CLIN-2006, Leuven, Belgium

Van Eynde F (2005) Part of Speech Tagging en Lemmatisering van het Corpus Gesproken Nederlands. K.U. Leuven, (in Dutch)

Vossen P (1998) EuroWordNet: A Multilingual Database with Lexical Semantic Networks. Kluwer Academic Publisher

Vossen P, Maks I, Segers R, van der Vliet H (2008) Integrating lexical units, synsets, and ontology in the cornetto database. In: Proceedings of LREC-2008, Marrakech, Morocco

Towards a Discourse-driven Taxonomic Inference Model

Piroska Lendvai

Abstract This chapter describes ongoing work, the goal of which is to create a discourse-driven inference model, as well as to construct resources using such a model. The data process consists of texts from two encyclopedias of the medical domain—stylistic properties characteristic of encyclopedia entries constitute the mechanisms underlying the inference model, such as layout-based features along-side with semantic (conceptual) document structuring. Three parts of the model are explained in detail, providing experimental results that are based on language processing techniques: (i) identifying taxonomic document structure by machine learning; (ii) discourse-driven construction of text–hypothesis pairs for examining types of textual entailment; (iii) semi-supervised harvesting of lexico-semantic patterns that connect medical concept types.

1 Introduction

The question answering (QA) system developed by the ROLAQUAD group[1] within the IMIX project is created to answer user queries about general medical information. It operates on the basis of a semantically annotated reference text collection that consists of parts of two medical encyclopedias published in The Netherlands, the Merck Manual (Berkow, 2000) and the Spectrum Medical Encyclopedia (Spectrum, 2003). The QA module first identifies the medical terms and the topic (e.g. cause, symptom, treatment, etc.) in the user's question, and subsequently matches these to the encyclopedia texts that are manually annotated on a word, sentence and paragraph level.

Piroska Lendvai
Research Institute for Linguistics, Hungarian Academy of Sciences, Budapest, Hungary, e-mail: piroska@nytud.hu

[1] http://ilk.uvt.nl/rolaquad

A small corpus of 67 user queries was collected from naive users by deploying the first version of the QA system, containing questions such as *"How do you find out whether you have familial hypercholestrolemia?"*. A straightforward strategy in QA for selecting documents that possibly contain the answer is to match the term identified in the question to words in the (section) titles of encyclopedia entries, in order to return a complete paragraph from an encyclopedia entry as the answer. It is possible, however, that the medical terms appearing in these questions are underspecified: nevertheless, these may yield many possible candidate answers, coming from semantically mutually exclusive sections of documents.

For example, to answer the question *"What are the symptoms of meningitis?"*, mapping 'meningitis' yields two different sections of the encyclopedia entry on 'meningitis', both describing symptoms: by examining them, it is evident that the first section is titled *Bacterial meningitis*, the second *Viral meningitis*. One of the sections starts with the sentence *"Bacterial meningitis is caused by bacteria, among others by meningococcus (Neisseria meningitidis)..."*, and the other, in a similar fashion, starts with the sentence *"Viral meningitis (lymphocytic meningitis), meningitis caused by a virus, probably occurs ...*[2]. After explaining the causes, both sections describe symptoms and treatment for the given subtype of meningitis. To alleviate the contradiction that more than one segment of text can be returned to the user as an answer, with different content, underspecified user queries need to be regarded as ambiguous, and a clarification question, for example, needs to be generated for the user to specify a narrower term (or, as a simpler solution, a warning needs to be generated about multiple variants of the term 'meningitis'). To enable such intelligent feedback within the QA system, it is useful to include a subroutine in the (offline or online) processing of the reference documents of the QA system, which identifies whether a document elaborates on subtypes of the title term.

The first part of the contribution explains the development of a machine learning approach to automatically find conceptual taxonomy in medical documents. The second and third parts consist of discourse-driven construction of text - hypothesis pairs for examining types of textual entailment, as well as proposing a semi-automatic way to extract lexico-semantic patterns that connect medical concept types, and integrate these into the generic semantic inference framework emerging in the language technology community, based on discourse-driven text segmentation.

The aim hereby is to work towards a discourse-driven taxonomic inference model, consisting of an *n*-tuple of domain terms, where the hypernym is linked to its co-hyponyms via a specific relation. Since these relations are domain specific, they need to be identified in a data-driven way, that captures the phenomenon that in principle a large variety (in any case much more than what is hard-coded in current semantic resources) of two or more terms can stand in a co-hyponym relation with respect to a third term, if evidence is found for this in the text. The data used in this study comes from two Dutch medical encyclopedias, but it is possible to process other encyclopedia texts using the model, and this should be the target of

[2] All sentences are translations of the Dutch original text.

future work. This chapter will argue that current methods for data-driven relation extraction (i.e. as described in the chapter by Bouma, Fahmi, and Mur, *Relation Extraction for Open and Closed Domain Question Answering*, this volume) specify fewer details of the relations than the discourse-driven model proposed. It is also argued that embedding such a model in a general semantic framework, such as semantic inference, is beneficial for formalising the complex phenomena being dealt with.

The model would for example hold tuples such as:

```
meningitis - is caused by - [] (bacterium; virus)
gallbladder disease - exists in - form (acute; chronic)
digestive tract - consists of - organ (tongue; esophagus; stomach)
jaundice - can be developed by - [] (baby; adult)
ganglion blockade - can take place via  - method (surgery; medication)
```

and so on. Note that 'tongue', 'esophagus', and 'stomach' are co-hyponyms in an *is_a* relationship with their hypernym 'organ', but are also co-hyponyms (or: taxonomic siblings) with respect to the term 'digestive tract', in a meronymic (i.e. part-whole) relationship. The discourse-driven taxonomic inference model aims to hold and promote especially such domain-specific, non-*is_a* relationships. Also note that the *is_a* hypernyms might not always become available from the model (their place is marked by the empty '[]' brackets), they are however often obtainable from general semantic resources with taxonomic or ontological structure, such as Wordnet, Wikipedia Categories, and domain ontologies.

In addition to the language processing exercises performed in each of the three subtasks, the model is developed in order to be applied to unstructured documents where fewer or no discourse-level cues are available (e.g., documents without explicit layout such as paragraphs or sections), in order to discover taxonomic structure in those and acquire new co-hyponyms linked to their coordinating term via a specified relation.

1.1 Conceptual Taxonomy

The method for recognising taxonomic structure in encyclopedia entries exploits manually assigned semantic annotations. Section 2.1 explains in detail that, annotations are assigned to three levels of text. The learning algorithm is trained on (strings of) words, as well as the manually assigned labels that encode medical concept types, as well as sentence- and section topic types, based on which it classifies whether semantic taxonomy is present between the text coming from two different sections of one encyclopedia entry. In the example document on meningitis, the subtypes 'viral meningitis' and 'bacterial meningitis' are described in separate sections, and the content of both sections are characterised with three manually assigned tags that in the given case label the section topics `cause`, `symptom`, and `treatment`. Therefore, these two sections are seen to describe co-hyponyms (also known as co-ordinate terms, or taxonomic siblings), because, judged by the assigned labels, the subtypes of the same disease share the same three semantic relations with the main term described by the document.

Based on labelled examples, a memory-based learning algorithm identifies whether two sections of a document are about taxonomic siblings or not. The implicit knowledge to be learned by the algorithm in this case is that there is semantic overlap (in terms of cause, symptom, and treatment) between these two sections of the document. This classification can be utilised to infer that the terms 'viral meningitis' and 'bacterial meningitis' are subtypes of the main term 'meningitis', *co_hyponyms*(bacterial meningitis; viral meningitis). This is in fact identifying the *is_a* relation between these narrower terms, and meningitis, the broader term:

```
(bacterial meningitis; viral meningitis) - is a - meningitis
```

The taxonomy can be reused in understanding and clarifying user questions to the QA system.

As said above, however, it is regarded as more important to extract the specific reason (i.e. the underlying knowledge of facts) that explains why the *is_a* relation holds. This is explained if the triplet can be created

```
meningitis - is caused by - [] (bacterium; virus)
```

based on the argumentative structure that is often reflected in the layout of encyclopedia entries, as well as by lexical patterns that can be harvested from these texts.

Note that in order to obtain the exact terms and relations (i.e. in the form of lemmas) that make up a tuple, some morphosyntactic processing needs to be interfaced with the model, e.g. for transforming the actually occurring string 'bacteria' into its singular form 'bacterium', linking the adjective 'bacterial' to the noun 'bacterium', etc. However this issue will not be dealt with, apart from referring to literature that discusses methods targeting this.

1.2 Discourse Structure

It is shown in several studies that thematic coherence impacts document structure, hence identifying discourse relations can facilitate access to semantic content; for an overview of relevant issues in the field of discourse parsing see (Péry-Woodley and Scott, 2006). In the case of stylistically guided documents, such as encyclopedia entries, pressure for brevity and clarity is superimposed on the creative interplay between form and factual content of narration. At the same time, as already seen, the entry often provides a definition of a term of which several subtypes exist and are thus separately described – in a sense, in a linguistically parallel fashion, even if in a structurally linear way – the use of tables, charts, bulleted lists and other visual means that are inherently well suited for representing hierarchically structured data is traditionally not favourable in dominantly textual media. A solution for this is to provide lexical, syntactic, and discourse cues to the reader for inferring the links, and the link types, between the related terms.

Linguistic constructions that are employed to convey structured information are for example syntactic coordination, apposition, and 'macro-propositions', such as pre-announcements of certain subtopics that are going to be addressed in the document, e.g. *"There exists an acute and a chronic form of gallbladder disease."*, *"There are several forms of examination of the stomach,"*, etc., as well as recurring lexical and syntactic elements ('chains') across the passages treating the subtopics, as in the excerpt from the encyclopedia entry on *Digestive tract*:

> The first organ is the tongue which is only present in the phylum Chordata. The second organ is the esophagus. ... The third organ is the stomach. ...

Such constructions serve to alleviate the readers' cognitive load required to make semantic inference about hierarchical and semantic relations (e.g., that *tongue*, *esophagus*, and *stomach* are co-hyponyms of the hypernym *digestive tract*, all three are *organs* of it, etc.). Most importantly, many encyclopedia articles treating co-hyponyms do *not* feature any of these linguistic means, but operate by activating real-world knowledge, therefore it is an empirical research issue how to trace back and formalise links between co-hyponyms and their hypernym. Once discourse relations – in the current study kept at the level of semantic overlap between pre-segmented passages of text (i.e., sections of a document) – are identified in a text, access to the semantic content is going to improve, as it can then proceed in a structured way.

1.3 Semantic Inference

In the past years, the notion of textual entailment (TE) has been introduced as a generic framework for applied semantic inference over texts. This framework aims to create the means for modelling language variability in an application independent manner (Dagan and Glickman, 2004). The motivation for creating a TE framework is that inferences used in various problems in natural language processing (e.g. QA, information extraction, summarisation) can be reduced to entailment as a phenomenon, so that the scattered tasks that all involve applied semantics will be represented within one, theory-neutral, framework. An overview of existing methods in TE is provided in the recent study of Androutsopoulos and Malakasiotis (2010), while the work of Hickl et al (2006) mentions how to collect additional entailment pairs. Experimental acquisition of entailment rules is described and evaluated among others in Szpektor et al (2007).

TE is a directional relation between two text fragments, Text (t) and Hypothesis (h): t entails h if humans reading t will infer that h is most likely true. For example, *X's purchase of Y* entails that *Y is owned by X*. The computational task then consists of generating h from t automatically. Alignment of shared variables between t and h plays an important role in this, whereas language and world knowledge are fed into the TE model using resources in the form of inference patterns (e.g. the example above) and rules (e.g. syntactic tree transformations)[3]. Parts of the model still await being discovered and assembled; importantly, recent work by the community focuses on integrating properties of discourse into TE, for instance resources dealing with co-reference resolution (Mirkin et al, 2010).

This study argues that discourse-level taxonomic inference can be seen as a restricted type of TE and that it is, therefore, possible to utilise discourse structure to automatically generate training data from documents, to benefit the TE framework. For example, by creating *t* - *h* pairs and the corresponding class, stating if the pair is a positive or negative example of entailment, and marking up the variables that motivate this class (i.e. the variables that are shared across *t* and *h*). In the second part of this study, the focus is on alignment of elements that are manifested in taxonomically structured encyclopedia entries. Based on the observations of the elements' appearance with respect to each other, it is argued that it is possible to create *t* - *h* pairs that resemble subtypes of TE, if parts of an encyclopedia entry featuring taxonomic structure are rearranged. Namely, an encyclopedia entry's introduction section emulates the inferred knowledge (that can be seen as the *h* in the TE framework), when it describes subtypes of some concept. The sections elaborating on the subtypes are joined together to form *t*.

Using the discourse-driven taxonomic inference model, from e.g. the example material on *digestive tract* in Section 1.2 one would be able to generate the *h*: 'The digestive tract (of the phylum Chordata) consists of three organs', based on the extracted tuple.

```
digestive tract - consists of - organ (tongue; esophagus; stomach).
```

1.4 Semi-supervised Harvesting of Lexico-semantic Patterns

The third component of the targeted taxonomic inference model consists of extracting patterns that combine lexical (i.e. term-level) and conceptual (i.e. ontological class-level) elements from encyclopedia texts. This is addressed by annotating lexical strings with supercategories of medical terms (i.e. abstract types of medical concepts), e.g. `disease_symptom`, `method_of_diagnosis`, or `treatment` (see the rightmost column of Table 1).

Harvesting takes place in a semi-supervised way: these manually assigned annotations of medical term types are used to 'mask' the actual lexical strings, which are subsequently processed applying an unsupervised technique, grammar induction. The collected co-occurrence patterns often capture lexical variation of relations, often in terms of verbal valency, e.g. '<advice> against <disease>', '<advice> specifically given to <person>'. The taxonomic inference model is to be utilised for extending and enriching the repository of relations between entities.

[3] see http://www.aclweb.org/aclwiki/index.php?title=Textual_Entailment_Portal

1.5 Related Research in Language Technology

Automatic construction of a domain ontology from texts is a heavily researched area in computational linguistics, where a number of advanced technologies have been created in the past decades, as surveyed in Buitelaar et al (2005). The (semi)-automatic extraction of hyponym-hypernym terms is a popular topic in the literature, initiated by the seminal paper of Hearst (Hearst, 1992), which is often addressed in the wider field of ontology construction (see e.g. McDowell and Cafarella, 2008). One work in this field that treats specific, hierarchical relations is Wang et al (2006), in which the relations are binary, term and relation recognition are not integrated, and the elements of the domain model are predefined.

The method to detect that two sections describe co-hyponym terms differs from traditional text-based approaches to ontology population, such as those using syntactic dependencies (Declerck and Vela, 2005), morpho-syntactic pattern matching, or heuristics on unstructured text (cf. Cimiano et al, 2005 and its references). The phenomenon that semantically related words tend to share similar syntactic contexts is often exploited in order to build taxonomical lexical resources, for example in the work of Van der Plas (2008), using these contexts to determine the semantic relatedness of words. Note that the work so far does not incorporate making use of syntactic information.

Kozareva et al (2009) show how such an extraction mechanism can be bootstrapped from texts on the web, building it in a Learning by Reading system as a general framework for this given language processing task. They however focus solely on *is_a* relationships, using a definition of *is_a* that allows a term (i.e., an instance of a concept) to have several *is_a* relationships at the same time. Their algorithm learns new hypernyms (superordinate terms) for a set of subordinate terms; e.g. given a set of animal instances such as *(dog, cat)*, it discovers new terms that are superordinate category names, e.g. 'mammal', 'pet', and so on. An important issue with respect to the type of relationships that are obtained should be noted. While it is possible to relate these superordinate category names to each other via the *is_a* relation, in fact a dog *is a* pet only under certain circumstances, thus in the ontology building world 'pet' would not be considered as a proper hypernym of 'dog'. Guarino and Welty (2002) establish that the notion of rigidity is of great importance for useful inferences: according to this, 'mammal' is a rigid concept, whereas 'pet' is not – it is a potentially applicable property of dog. The distinction is vitally important to make in order to build structured resources that allow for semantic inference, justifying the validity of this approach for promoting domain-specific relations by the taxonomic inference model. It is argued that it would be useful to constrain such *is_a* relationships, or replace them by other relations, such as, for example, a dog *can be kept* as a pet – which is in effect what this work is targeting.

With respect to studies utilising discourse structure or layout, this corpus exhibits less hierarchical structure than scientific or technical texts, utilised e.g. by Makagonov et al (2005) for inferring domain ontology. Our approach also differs from semantic classification of medical document segments described in Cho et al

(2003), since taxonomic relations are discovered between predefined segments of a document.

2 Exploratory Data

Our reference corpus contains parts of the Merck Manual and the Spectrum Medical Encyclopedia. The corpus was automatically segmented on four levels using a tokeniser and exploiting existing XML markup, into words, sentences, sections, and documents. The documents, each representing an encyclopedia entry, were manually annotated based on a protocol for a set of medical concept types (spanning one or more words), topics (spanning a sentence), and section topics (spanning a section). The three tagsets are detailed in Table 1. The corpus contains 3,178 documents. There are 6,582 document (sub)sections; 1,716 documents (54% of the corpus) consist of only a single section. The average number of sections present in a document is 2.1. The average document length is 6.3 sentences.

2.1 Semantic Annotation Types

In some of the multi-section documents, subterms of the header term in the section titles can be identified, detecting these either manually, or automatically, drawing on the (manually assigned) section type annotations. It is assumed that a section type that is applicable to two or more sections of a document indicates semantic overlap between the given text segments. With this approach 128 documents can be identified that exhibit conceptual taxonomy. It is a small fraction of the total document collection; encyclopedia texts that recently became available (e.g. Wikipedia) would yield much more data.

Validating the results of the identification requires close inspection of the text structure and content. In particular since this corpus consists of two encyclopedias that are clearly constructed along different editorial lines. By manually checking the identified documents, it is indeed possible to find taxonomy of domain terms in these articles. For example, co-hyponyms that indicate types of ailments of an organ (e.g. malignant vs. benevolent tumor in some organ) can be observed, coordinated terms that are types of treatments (e.g. internal or external application of a medicament), co-hyponyms of medical procedures or physiological processes (e.g. delivery by cesarean section versus helped by vacuum extractor), and the like.

There are 15 different types of medical topics annotated in the corpus on the level of document sections; this set was defined by examining the main types of domain semantics in the reference text. Sometimes a section cannot be labelled by any of these tags, sometimes more than one tag applied to a section. Subsections are unlabelled. For example, the entry titled *Carpal tunnel syndrome* has four sections, labelled as `definition`, `symptoms`, `cause`, and `treatment`,

Table 1 Three semantic annotation tagsets applied to the reference corpus: document section topic, sentence topic, and medical term supercategory.

Document section topic	Sentence topic	Concept supercategory
Applications	Causes	Bodily_function
Cause	Definition_of	Body_part
Consequences	Diagnoses	Disease
Contamination	Is_a_kind_of	disease_feature
Definition	Is_property_of	Disease_symptom
Diagnosis	Is_side_effect_of	Duration
Diseases	Is_similar_to	Method_of_diagnosis
First_aid	Is_symptom_of	Micro-organism
Forms	Is_synonym_of	Person
Incidence	Is_transferred_by	Person_feature
Methods	Occurs_in	Treatment
Prevention	Prevents	Treatment_feature
Side_effects	Treats	
Symptoms		
Treatment		

respectively. Sections are also augmented with their original Dutch encyclopedia section titles (if any), these may correspond to the topic annotations (e.g. in this case 'Symptomen', *symptoms*, 'Oorzaak', *cause*, and 'Behandeling', *treatment*); the first section of multi-section documents usually is a general introductory section, and has no section title on its own.

An example of a more complex structure is the document about *Sterilisation*, which has three sections annotated as (1) `definition`, (2) `applications, consequences, definition, method`, and (3) `applications, consequences, definition, method`. The second section's title is 'Bij de man' (i.e., *Of men*), whereas the third section is titled 'Bij de vrouw' (*Of women*). This document clearly encodes conceptual taxonomy, which may not be straightforward to discover automatically, since it is only implicitly marked by the repeated string of words ("bij de") and ontologically related concepts ('man' and 'vrouw') in the section titles, as well as in the content of the sections. The task is to investigate how such conceptual taxonomy can be identified on the basis of the overlap in semantic annotations available to us on the concept-, sentence-, and section level.

On the sentence level thirteen semantic topic categories are annotated. It is possible that several different topics are assigned to one sentence. On the concept level twelve semantic supercategories are assigned to domain entities, such as `disease` or `body_part`.

When marked up with the three tag sets, the second sentence of the article on *Hersenvliesontsteking* (i.e. *meningitis*) is annotated as follows:

`<SECTION: cause,definition>` `<TOPIC: definition_of,is_transferred_by,causes>` *The disease can be caused by several types of* `<CONCEPT: microorganism>` *bacteria* `</CONCEPT: micro-`

organism> *and* <CONCEPT: microorganism> *viruses* </CONCEPT: microorganism>. </TOPIC> </SECTION>

3 Machine Learning of Taxonomy Identification

Identifying and extracting components of conceptual taxonomy has been shown to have multiple facets. It was chosen to implement conceptual taxonomy detection as a supervised classification task, drawing on the annotated corpus. The exploratory experiments are designed in a bottom-up way: the first step is to automatically learn whether two sections of a document talk about concepts that are taxonomic siblings, or not. In other words, each document section is paired with every other section of the same document, and the pairs are classified as positive or negative instances of taxonomic siblings (co-hyponyms), regardless of the semantic aspect(s) the siblings would share. A memory-based learning algorithm[4] is used for this task. Using its default parameters, this algorithm assigns either the 'positive' or the 'negative' class to a test instance (i.e. two sections of a document), on the basis of the class of the most similar training example (i.e. pairs of document sections labelled as taxonomic siblings or not) it has seen.

3.1 Feature Construction

The learning algorithm draws on a binary vector representing overlap between two sections of a document in terms of an unordered bag-of-words (a lexicon of 5,421 bits), as well as on overlap between the set of concept supercategories, and the set of sentence topics (see Table 1) e.g., if both sections have the word 'meningitis' in them, the bit representing this word is set to 2; or if only one of the sections has the concept `body_part` annotated in it, the bit representing this concept is set to 1; or if no sentence in any of the sections is annotated with the topic `is_side_effect_of`, this bit is set to 0, and so on.

3.2 Experimental set-up

The two encyclopedias annotated differ from each other, for example, in how well they are structured, the consistency of section titles and the length of the articles. To capture the general suitability of this approach, separate experiments are performed on data from the Spectrum Encyclopedia (that consists of highly-structured documents according to a general scheme) and on the Merck Manual

[4] TiMBL, release 5.1. `http://ilk.uvt.nl/timbl`

(that is built less consistently. Its documents bear more resemblance to descriptive texts, where content steers structure). After generating section pairs, classification experiments on 697 Spectrum instances are run (174 positive, 523 negative), and 210 instances from Merck (49 positive, 161 negative).

Since the data is small, a leave-one-out testing method was performed: the whole dataset is used as training data except for one data point, which is afterwards used for testing. Evaluative metrics are overall accuracy, micro F-score (measured over all instances), macro F-score (measured over both classes), whereas for both classes precision, recall, and F-score is calculated. The focus of interest is the F-score as measured over the positive class, since this figure characterises how well taxonomic siblings are able to be identified, which then allows for inducing components of the discourse-driven inference model.

3.3 Results

The experimental results are shown in Table 2. On both data sets, classification yields the best scores when the two sections are represented in terms of the overlap between annotated sentence topics. Using this information, the learning algorithm successfully detects sections that describe taxonomic co-hyponyms with 60 points F-score on data coming from the Merck Manual, and 77 points F-score on data coming from the Spectrum Encyclopedia. The type of concept supercategories that can be found in a section also adds important information to guessing a taxonomic relation, but on a much smaller scale than sentence topics. Overlap between words in the two sections gives the least cues to the learning algorithm.

Combining the three feature types improves the algorithm's performance only when sentence topics and concept supercategories are combined; this feature set performs equally well on Spectrum data (a 78-point F-score) as sentence topics alone.

In general, and as expected, the scores are higher on Spectrum data, whereas those on Merck can be seen to represent the situation of inducing taxonomy from relatively free-form texts. Precision and recall on both positive and negative classes are quite balanced, which means that the algorithm is able to treat both classes similarly, despite that positive classes are in the minority in both datasets.

Certainly, the current results can be employed as baselines in more elaborate experimental implementations of the approach (e.g. testing classifiers other than the memory-based learner, employing predicted instead of annotated semantic features, employing morphosyntactic features and n-grams, etc.). In the taxonomy inference model however, the interest goes out to identifying the type of semantic topic that gives rise to such taxonomy (e.g. both subtypes of meningitis are described in terms of their cause, symptoms, and treatment), in the form of a lexicalised relation.

Table 2 Classification results of conceptual taxonomy between two sections of a medical document, based on leave-one-out experiments with memory-based learning, using varying feature representation, in terms of several evaluation metrics: overall accuracy, micro (FmI) and macro (FmA) F-score, as well as precision, recall, and F-score on the positive and negative classes.

Collection	Features	Overall scores			+ class			− class		
		Acc	FmI	FmA	Prec	Rec	F	Prec	Rec	F
Spectrum	BoW	57	58	44	16	17	17	72	70	71
	Concepts	75	75	67	51	49	50	83	84	84
	Topics	88	88	85	78	76	77	92	93	92
	Concepts+topics	89	89	85	79	76	78	92	93	93
	BoW+concepts	59	59	46	19	20	20	73	71	72
	BoW+topics	60	61	48	22	23	22	74	73	73
	BoW+concepts+topics	60	61	48	23	24	23	74	73	23
Merck	BoW	67	66	52	27	24	26	78	80	79
	Concepts	69	69	56	33	33	33	80	80	80
	Topics	79	80	73	55	65	60	89	84	86
	Concepts+topics	79	79	70	55	53	54	86	87	86
	BoW+concepts	69	68	53	29	24	27	78	82	80
	BoW+topics	68	67	53	29	24	26	78	81	80
	BoW+concepts+topics	69	68	53	29	24	27	78	82	80

4 The Taxonomy Inference Model and Textual Entailment

Representing specific relations between domain entities can be utilised in the construction of an ontological model that in turn is an important asset of knowledge-based applications that e.g. perform advanced information access. The work targets the detection of domain-specific lexico-semantic relations (sometimes called 'associative relations') as well as hypernymy and meronymy relations, and models the involved elements (that are typically domain-specific terms) as a taxonomy inference tuple.

A taxonomy inference model consists of an n-tuple of terms, two or more of them linked by a coordinate relationship (i.e. they are siblings), while each of these is linked by either hypernymy or meronymy or an associative relation to the common coordinating term (i.e. the mother node). The relations between mother and child nodes are governed by a certain semantic property of the mother node, which in the case of meronymic or hypernymic relation is best expressed by a noun phrase – for example, in the medical domain *method* of a treatment, *phase* of a process, *form* of a disease –, or, in the case of domain-specific relation, by a verb: such as something *occurs_in* a body part, *attacks* a microorganism, *causes* an illness, and so on. The coordinating relation can typically be induced from background (domain) knowledge, but often remains implicit in the text, making it difficult to automatically harvest it by computational mechanisms. While it is non-trivial to retrieve such relations from texts, especially for co-hyponymy cases, they are driven by an extremely productive mechanism of human cognitive inference.

The goal is to investigate the linking of coordinate terms via a specific type of relation, inferrable from a document collection. Instead of measuring the utility of sentence alignment based on standard, automatically obtained syntactic (dependency, POS) and semantic (cosine, lexical semantic databases) similarity cues for this task, the focus here is on the role discourse patterns can play in this process.

The complexity of phenomena underlying taxonomy inference is illustrated by the following entry excerpt (the document and section titles are set in italics).

> *Jaundice*, or icterus, is a condition whereby a yellowish discolouration of the skin, the mucous membranes, and whites of the eyes appears. This is caused by increased levels of the gall pigment bilirubin in the blood serum. ...
> *Adults* Adults can develop jaundice by three means. By an interruption to the drainage of bile in the biliary system (e.g. due to a gallstone or a tumor). ... By diseases of the liver (e.g. hepatitis). The liver's ability to metabolise and excrete bilirubin is reduced, leading to a buildup in the blood. ... By an increased rate of bilirubine production. ...
> *Babies* Babies can develop a sort of jaundice (icterus neonatorum) shortly after birth as a consequence of relatively increased breakdown of red blood cells ... .

In this entry, the onsets of the two sections exhibit some syntactic and lexical parallelism; after establishing this overlap, the aim is to extract the relation of the main terms of each of these sections (of *adults* and *babies*, i.e., of the co-hyponyms) with respect to the coordinating term (i.e., the entry title *Jaundice*). Spotting lexical overlap and using wildcards yields the expression *'X can develop Y'*, and it is assumed that the (verb) phrase between the two variables instantiates the domain-specific relation shared by the coordinate terms in connection to the term *jaundice*.

Note that this entry does not contain a separate passage of text that can be designated as the hypothesis (of semantic inference) of the above relation, thus the lexical string is fallen back on (i.e., the verb phrase) when extracting the relation. Also note that taxonomy inference can be embedded: the syntactic similarity among three sentences in the *Adults* subsection is observed too. The coordinating hypernymy relation is worded by the very first sentence of this section (hence regarded as the *inference hypothesis*), yielding the expression *'develop Y by (NUM) means'* – which is in fact a string that can be regarded as the hypothesis of the inference relation in the tuple

```
jaundice - can be developed by - means (interruption to the drainage of
bile; diseases of the liver; increased rate of bilirubine production).
```

In the current section, how this inference type relates to textual entailment is investigated. That the thematic coherence makes an impact on document structure has been observed, which is utilisable for information extraction purposes, the aim is to model taxonomic relations between entities, as they appear in context, as a special type of TE. The goal is to create entailment rules associated with a co-dependent 'contrast/elaboration' phenomenon.

Interestingly, in many cases it is possible to create text – hypothesis pairs from the encyclopedia documents used, again driven by shallow discourse structure. For example, the passage created from concatenating the two sections of the *Jaundice* entry can be regarded which elaborate on the co-hyponyms *Adults* and *Babies* as *t*, and generate *h* from the onsets of these two sections (as well as insert the coordinating element 'and' between them). The result is the following text – hypothesis pair:

Text 1

Adults can develop jaundice by three means. By an interruption to the drainage of bile in the biliary system (e.g. due to a gallstone or a tumor). ... By diseases of the liver (e.g. hepatitis). The liver's ability to metabolise and excrete bilirubin is reduced, leading to a buildup in the blood. ... By an increased rate of bilirubine production. ... Babies can develop a sort of jaundice (icterus neonatorum) shortly after birth as a consequence of relatively increased breakdown of red blood cells ... .

Hypothesis 1

Adults can develop jaundice (and) Babies can develop a sort of jaundice.

The pair that has been created illustrates a very simple form of TE, since all variables (in fact all words) in *h* can be directly identified in *t* as well. A more complex case is represented by the encyclopedia entry on *Ganglion blockade*, see Figure 1.

Fig. 1 Encyclopedia entry on *Ganglion blockade*. The mother node of the inference tuple is marked in boldface, variables that are synonymic/hyponymic terms are marked up in shades across introductory part and thematic sections. Coordinate terms (co-hyponyms) are *surgery* and *medication*.

Ganglion blockade

Ganglion blockade can take place via surgery or medication.

The surgical method is performed to increase the circulation and blood supply of the tissue that is in correspondence with the ganglion; impulses transmitted by nerves to narrow blood vessels are interrupted (sympathectomy). ...

Systematic treatment by ganglion blockers used to be practiced to remedy high blood pressure.

The mother node of the inference tuple extractable from this text is *Ganglion blockade*, the coordinate terms (co-hyponyms) are *surgery* and *medication*. The introductory section is the 'macro-proposition' that would correspond to *h* in the TE framework, and the two 'thematic' sections on the two different kinds of methods correspond to the *t* part in the entailment set-up. The tuple to be extracted from this entry is

```
ganglion blockade - can take place via - method (surgery; medication).
```

Interestingly, it is seen that the variables driving the semantic inference in this example are expressed by means of synonyms and hyponyms across the introductory part and the thematic sections (contrary to the entry on *jaundice*). These are marked up in shades in Figure 1. The strings (VPs) *is performed* and *used to be practised* are synonymous with *can take place*. *surgical method* is a term variant of *surgery*, and *treatment with ganglion blockers* is a narrower term of *medication*. So in fact the same inference tuple can be recreated in a number of variants, for example

```
ganglion blockade - is performed via - method (surgical; treatment with
ganglion blockers)
```

and so on. However, instead of generating such variants, it is preferable to follow best practices with respect to representing synonyms and the like of term variants (see also the recent work of Declerck and Lendvai (2010) on linguistic representation of terms in ontological resources). The issue of how such best practices would affect the resources generated for the TE framework (e.g. repositories of syntactic mapping rules, paraphrases, and the like) requires thorough investigation and unfortunately lies out of the scope of the present study.

Given the observations of this study, discourse-level taxonomic inference can be assumed to be seen as a restricted type of textual entailment. The introduction section of encyclopedia entries with taxonomical structure often emulates inferred knowledge (*h*), which is about subtypes of some entity or phenomenon. Specific discourse relations hold between the heuristically designated document parts: *Elaboration* between Introductory part and Section1 respectively Section2 (3, etc.), as well as *Contrast* between Section1 and Section2 (3, etc.). In these cases, it is possible to generate positive examples of entailment by joining Section1 and Section2 (3, etc.) into *t*, and regard the introductory part of the document as *h*. Applying methods of the TE framework to the $t - h$ pair would result in specifying rules for the specific taxonomic inference relation between these passages of text, and would further exemplify the various mechanisms underlying semantic inference in terms of syntax (alignment, transformations), lexicon (taxonomic relatedness), and semantic equivalence (paraphrased constituents).

Difficulties in implementing this occur on several levels. Most importantly, pieces of information (entities, relations) are not consistently reoccurring across the Contrast and/or Elaboration sections. Furthermore, some subsections do describe coordinate terms, but these are not necessarily subtypes of the head entity. Finally, the lack of discourse markers further hinders formalisation. To illustrate this complexity, in Figure 2 the encyclopedia entry on *Lactic acid* is shown.

Lactic acid

Lactic acid is an organic acid that emerges via fermentation, occurring in two optically active forms (right-rotating and left-rotating), among others in sour milk.

In foodstuff

Fermented by bacteria, lactic acid can be present in all kinds of foodstuff (sauerkraut, buttermilk, yoghurt, wine), depending on the type of bacteria, it can either be of a right-rotating or a left-rotating form, whereby the right-rotating form is supposedly more beneficial for health than the left-rotating.

In muscles

Left-rotating lactic acid is produced in the muscles due to a shortage of oxygen.

Fig. 2 Encyclopedia entry on *Lactic acid*. Related variables are marked up in shades and underlining across introductory part and thematic sections.

In future work one would like to focus on identifying (lexico-syntactic) entailment rules associated with the co-dependent Contrast/Elaboration phenomenon, by examining patterns occurring across sentences and learning about the context of related terms. The goal would be to produce evidence to the TE claim that application inferences (here: taxonomic structure reflected by discourse) can be reduced to (partial) entailment.

5 Extraction of Patterns Involving Medical Concept Types

The core component of the third method utilises unsupervised structure induction to create clusters of (strings of) words from the sentences in a document. Grammar inference systems are able to bootstrap a structure made up of syntactic-like constituents, without actually having any syntactic or semantic knowledge (i.e., without requiring annotated examples) or knowing the type of language they are applied to. The strength of using this method is thus its general applicability and no need for linguistic resources; its utility is exemplified in e.g. the study of Katrenko and Adriaans (2006).

Unsupervised Grammar Induction

From the variety of grammar induction systems, the Alignment-Based Learning (ABL) algorithm (Van Zaanen, 2001)[5] was chosen. The ABL algorithm finds a grammar that underlies a corpus of plain text sentences. The first cycle of ABL aligns the sentences in the input corpus. The procedure can be likened to bracketing the words in one sentence by comparing it to every other sentence in the corpus, on the basis of string similarity metrics. If two or more sentences contain identical (sequences of) words, the non-overlapping parts of the sentences are hypothesised as grammatical constituents of the same type, because they could in principle be interchanged in the given context. The grammar induced by such inference systems can be visualised in terms of production rules from symbolic non-terminals to terminals.

5.1 Masking

The ABL algorithm draws on string similarity. To optimise its working, certain groups of words to a symbolic token are collapsed; this procedure is called *masking*. In the current study, masking was manually done by assigning medical concept types to the relevant string of words, as explained in Section 2.1. Masking introduces more homogeneity into a document and thus facilitates the extraction of higher-level patterns, so that e.g. in the study the ABL algorithm induces a grammar that focuses on verbs in relation to general character categories, rather than having to calculate string overlap between the actual tokens that are now masked by concept

[5] Available from http://www.ics.mq.edu.au/~menno/research/software/abl

supercategories. Masking is illustrated by a few random examples from the data (translations are in italics).

<disease> of <disease> komt het meest voor in het <body_part> en op de <body_part> en de <body_part>.
<disease> or <disease> occurs most frequently in the <body_part> and on the <body_-part> and the <body_part>.
Als bij iemand <disease> wordt vastgesteld, wordt in zijn omgeving een <method_of_-diagnosis> gedaan.
If someone is diagnosed with <disease>, a <method_of_diagnosis> is carried out in his environment.
Hierbij wordt vaak een combinatie van verschillende soorten <treatment> toegepast.
Hereby often a combination of different kinds of <treatment> is applied.

Automatically Identifying Modifier-dependency Relations

In the constituent hypotheses of ABL, the simple heuristics of this study identify terminals that are able to take dependents, in a way similar to that of a head-modifier relation. The method was successfully applied to extract patterns from textual databases; for a detailed description see Lendvai (2008).

According to the definition used for this study, the 'modifier' may constitute both the immediate left and right context of the 'head'. Note that the terminology is ours, ABL does not provide any information in terms of traditionally established grammatical relations in the induced grammar. The modifier is retrieved at the phrasal (not the word) level from the corpus. The goal is to obtain strings that repetitively occur in the data, i.e. to capture semantic building blocks of the encyclopedia – without actually incorporating grammatical information in the extraction process.

Post-processing

ABL is run on the masked encyclopedia texts, followed by the heuristics. The output is then post-processed, filtering strings where at least one concept type and one verb are present. So, for example, the string extracted by ABL *always <disease_symptom>* is discarded, because it does not feature any verb, but the string *however also develop <duration>*is kept. Note that many of the strings do not feature syntactically correct phrase boundaries, since ABL generates them based solely on the (rather limited) data.

5.2 Experimental Results

Using this approach, several patterns are able to be extracted involving medical concept supercategories. Relevant strings are manually selected, based on the 'looks-good-to-me' approach, which is common in evaluating the output of un-supervised learning. Note that the approach is semi-supervised: it consists of an unsupervised extraction procedure on concept labels that are manually annotated. The obtained relation patterns are not annotated, only manually clustered for observation purposes. The selected strings are grouped according to the amount of concept supercategories the extracted verbal chunk takes. The list of strings extracted is provided below for illustration purposes.

True Unary Relations

Instances of this group appear to be valid grammatical chunks, ready to be stored as a string corresponding to a specific relation. A second concept supercategory is not required for grammatical completeness.

- <disease> wordt behandeld door een arts
<disease> is treated by a doctor
- <disease_symptom> gaat bij rust meestal over
<disease_symptom> often disappears while resting
- <disease_symptom> verdwenen zijn
<disease_symptom> are gone
- <treatment> niet mogelijk of zinvol is
<treatment> is not possible or meaningful
- deze <microorganism> komt in twee vormen
this <microorganism> has two forms

False unary relations

In these extracted phrases, an obligatory argument is missing, which is typically represented by another medical concept.

- <disease> kunnen veroorzaakt worden door sommige
<disease> can be caused by some
- <disease_symptom> ontstaat
<disease_symptom> appears
- <body_part> aangetroffen kan worden
<body_part> can be affected
- gaan gepaard met <disease>
co-occur with <disease>
- <body_part> binnendringt
penetrates into <body_part>

Binary relations

Instances of this group involve two medical concepts. Such strings are found quite rarely, as most of the extracted strings are ungrammatical.

- waarbij <disease_symptom> op de <body_part> ontstaan
whereby <disease_symptom> occurs on the <body_part>
- als gevolg van het <bodily_function> van asbestvezels die diep in de kleinste <body_-part> of zelfs
as a consequence of the <bodily_function> of asbest fibers deep into the smallest

<body_part> or even
- zodat de <body_part> - <body_part> - weer een kern met een <body_part> bezit
so that the <body_part> - <body_part> - again possesses a nucleus with a <body_part>
- tweede <treatment> aangeraden voor <person>
second <treatment> is advised for <person>

In general, it is observed that verbs occur less frequently in the substrings created by the method – it might be the case that masking covers certain verb phrases (for example by the concepts `treatment` or `method of diagnosis`). It is also the case that verbs that occur with a high frequency in the phrases are typically auxiliaries and other, domain-neutral verbs. This is probably due to the effect of stylistic properties of the corpus and the use of a specific vocabulary. It seems that it is problematic for the grammar induction method to align strings on domain-specific verbs – these might be typically unique, occurring thus with a low frequency across encyclopedia entries. This suggests that stylistic properties (extensive use of passive constructions, nominalisation, adjectival modification) of language used in medical texts impact the grammar of these texts in such a way that they become suboptimal for processing by the given method.

6 Closing

This chapter has discussed three natural language processing approaches, that address various facets of a discourse-driven taxonomy inference model proposed for knowledge discovery from encyclopedia texts. It is important to note that the methods shown in this study are transferable to new texts. Departure from the two medical encyclopedia, and the application of the techniques to new data will not only expand the list of *n*-tuples with new terms and relations, but will also update it. For example, processing the entry on 'Meningitis' in Wikipedia, will result in the following, updated *n*-tuples:

```
meningitis - can by caused by - microorganism (bacterium; virus)
meningitis - can by caused by - infection (bacterial; viral; aseptic; parasitic)
```

The results need to be manually evaluated. The system developed by the discourse-driven taxonomy inference model is eventually applied to the full IMIX corpus of Dutch medical encyclopedia texts to process single-section (i.e., unstructured) documents and automatically detect new coordinate terms and link them to the mother node by domain-specific relations. The obtained terms are matched to Dutch terminological resources available in the medical domain[6].

[6] `http://taalunieversum.org/taal/terminologie/medisch/`

References

Androutsopoulos I and Malakasiotis P (2010) A Survey of Paraphrasing and Textual Entailment Methods. Journal of Artificial Intelligence Research, vol. 38, pp 135–187

Berkow R (ed) (2000) Merck Manual Medisch handboek. Bohn Stafleu Van Loghum

Buitelaar P, Cimiano P, Magnini B (eds) (2005) Ontology learning from text: Methods, evaluation and applications. IOS Press

Cho P, Taira R, Kangarloo H (2003) Automatic segmentation of medical reports. In: Proc. of AMIA Symposium, pp 155–159

Cimiano P, Pivk A, Schmidt-Thieme L, Staab S (2005) Learning taxonomic relations from heterogeneous sources of evidence. In: Buitelaar P, Magnini B, Cimiano P (eds) Ontology Learning from Text: Methods, Applications, Evaluation, IOS Verlag

Dagan I, Glickman O (2004) Probabilistic textual entailment: Generic applied modeling of language variability. In: PASCAL Workshop on Learning Methods for Text Understanding and Mining, Grenoble

Declerck T, Lendvai P (2010) Towards a standardized linguistic annotation of the textual content of labels in knowledge representation systems. In: Calzolari N, Choukri K, Maegaard B, Mariani J, Odijk J, Piperidis S, Rosner M, Tapias D (eds) Proceedings of the Seventh conference on International Language Resources and Evaluation (LREC'10), European Language Resources Association (ELRA), Valletta, Malta

Declerck T, Vela M (2005) Linguistic dependencies as a basis for the extraction of semantic relations. In: Wroe C, Gaizauskas R, Blaschke C (eds) ECCB'05 Workshop on Biomedical Ontologies and Text Processing

Guarino N, Welty C (2002) Evaluating ontological decisions with OntoClean. Commun ACM 45(2):61–65

Hearst M (1992) Automatic acquisition of hyponyms from large text corpora. In: Proc. of COLING

Hickl A, Williams J, Bensley J, Roberts K, Rink B, and Shi Y (2006) Recognizing Textual Entailment with LCC's Groundhog System. In: Proc. of the Second PASCAL Recognizing Textual Entailment Challenge. Venice, Italy.

Katrenko S, Adriaans P (2006) Grammatical inference in practice: A case study in the biomedical domain. In: Grammatical Inference: Algorithms and Applications, vol 4201, Springer, pp 188–200

Kozareva Z, Hovy E, Riloff E (2009) Learning and evaluating the content and the structure of a term taxonomy. In: Proc. of AAAI 2009 spring symposium "Learning by Reading and Learning to Read"

Lendvai P (2008) Alignment-based expansion of textual database fields. In: Gelbukh A (ed) CICLing 2008. LNCS, vol. 4919, Springer Berlin / Heidelberg

Makagonov P, Figueroa A, Sboychakov K, Gelbukh A (2005) Learning a domain ontology from hierarchically structured texts. In: Proc. of ICML workshop on

Learning and Extending Lexical Ontologies by using Machine Learning Methods, pp 50–57

McDowell L, Cafarella M (2008) Ontology-driven, unsupervised instance population. Journal of Web Semantics 6(3)

Mirkin S, Dagan I, Padó S (2010) Assessing the role of discourse references in entailment inference. In: Proc. of ACL

Péry-Woodley MP, Scott D (2006) Introduction to the special issue on computational approaches to discourse and document structure. Traitment Automatique des Langues 47(2)

Van der Plas L (2008) Automatic lexico-semantic acquisition for question answering. PhD thesis, Groningen

Spectrum (2003) Winkler Prins Medische Encyclopedie. Spectrum

Szpektor I, Shnarch R and Dagan I (2007) Instance-based Evaluation of Entailment Rule Acquisition. In: Proc. of ACL

Wang T, Li Y, Bontcheva K, Cunningham H, Wang J (2006) Automatic extraction of hierarchical relations from text. In: Lecture Notes in Computer Science, Springer

Van Zaanen M (2001) Bootstrapping structure into language: Alignment-based learning. PhD thesis, School of Computing, University of Leeds, UK

Part V
Epilogue

IMIX: Good Questions, Promising Answers

Eduard Hovy, Jon Oberlander and Norbert Reithinger

Abstract The IMIX Programme was designed as a coordinated framework for addressing the difficult problems that arise in integrated multimedia information delivery. The programme, carried out by research teams in the Netherlands, funded by the Netherlands Organisation for Scientific Research, was very ambitious, combining research on automatic speech recognition within the context of multimodal interaction, dialogue management and reasoning, information presentation in multimodal systems, and information extraction. It managed to strengthen multidisciplinary collaboration, knowledge transfer between academia and industry, and the position of the Dutch language in the information society. The IMIX Demonstrator improved the visibility of language and speech technology as enabler of advanced information services. A follow-up research programme should include both a common framework, containing data, tasks, and evaluation, as well as a serious human factors evaluation component. Interactions and synergies between participants from areas that approach human communications from different angles should be actively encouraged.

1 The Legacy of the IMIX Programme

The IMIX Programme[1] documented in this book brought into the Dutch research sphere a coordinated framework for addressing the difficult problems that arise in

Eduard Hovy
USC Information Sciences Institute, Marina del Rey, CA, USA, e-mail: hovy@isi.edu

Jon Oberlander
University of Edinburgh, Edinburgh, Scotland, e-mail: j.oberlander@ed.ac.uk

Norbert Reithinger
DFKI Berlin, Berlin, Germany, e-mail: norbert.reithinger@dfki.de

[1] http://www.nwo.nl/imix

integrated multimedia information delivery.[2] Even at this early date, the programme leaves a legacy consisting of several parts. Firstly, and most centrally, it has firmly established in the national research scene the importance of making sure that various strands of language and multimedia technology research—information extraction, Question Answering (QA),speech recognition, dialogue management, etc.—are all part of a larger, integrated, picture. Secondly, the programme helped create a new generation of researchers, most of whom are now working on other (related) projects within the Netherlands and abroad, and who know one another's problems, methods of approach, and to some degree technical languages. Thirdly, and connected to the other two, the IMIX perspective already has a follow-up of sorts in the Dutch Language Union's STEVIN Programme[3], and will continue to inform research carried out in the Netherlands in ways rather different to most other countries (including the US, UK, Germany, France, Canada, Japan, and China), where funded programmes supporting this integrated perspective are considerably less apparent.

2 Evaluation of the IMIX Programme Work

The IMIX Programme had a very clear and logical set of goals, structure, and set of funded projects, in which good work was done, and of which the results were communicated well. The various dimensions were usefully and interestingly integrated. It was a very ambitious programme, with a remarkably successful performance leading to interesting results and an outstanding number of publications, given the limited financial resources. The programme also led to the creation of a new network of young talented researchers in the Dutch, as well as the international Human Language Technology field. Another similar follow-on research programme should certainly be proposed, after consideration of the comments on the strong and weaker points of the original programme, and the suggestions for possible new proposals.

The early creation of functional specifications of a common demonstrator certainly helped enable some integration and correspondence across the IMIX projects, and sufficiently served the goal of confronting the different researchers with various practical problems while integrating their individual modules into a common working Demonstrator system. For practical reasons, but mainly because of the limited budget of the IMIX Programme, relatively little funding was available for complete Demonstrator development, as a result of which the performance could not reach the level of a full prototype.

Unfortunately, there was little transfer of actual IMIX technology to industry. Long experience has shown that knowledge transfer from research to industry is very difficult to accomplish, and that technology transfer almost always requires

[2] This chapter is written by members of the international expert review panel of the IMIX Programme. The final review panel consisted of the authors of this chapter and Professor Steve Young (University of Cambridge).

[3] `http://stevintst.org`

human transfer as well to effectively carry over the expertise required. In this regard, the IMIX Programme resulted in one person working full-time in industry. Two others have worked as consultants, and several students and programmers from QADR are working in a QA start-up company.

However, the IMIX technology has seen some reuse. Several software components have been used in other projects, both within IMIX and outside the programme. The fact that IMIX helped to bridge the gap between the start and the funding of STEVIN was a hugely positive result.

The organisational structure of the IMIX Programme (steering group, programme committee, programme co-ordinator, technical integration, and programme office) was rather more elaborate than is typical in the US and Asia for government-funded research programmes of comparable size. Within IMIX, the overall organisation and management of the programmes were carried out very effectively, and to the Review Committee no problems were apparent. The loss and non-replacement of the co-coordinator required a more self-organising management by the various partners, which worked out remarkably well. The main drawback was that the Demonstrator, which was intended to focus the efforts and spur technical collaboration, did not retain the main focus of all groups.

2.1 Technical Evaluation

The IMIX Programme was very ambitious:

(i) Automatic speech recognition within the context of multimodal interaction;
(ii) Dialogue management and reasoning;
(iii) Information presentation in multimodal systems in natural language;
(iv) Information extraction.

It was simply impossible for a programme of this limited financial size and short duration to adequately cover the four thematic priorities in a balanced way. Given the realities of budget constraints, the research groups adapted the work plan very well to the requirements of the actual projects and the various external influences, including the departure of the coordinator, and the changing role of the Demonstrator. Their results and progress are well documented in this book, as well as in referred publications in leading journals and conferences.

When each project is contextualised, it is sometimes possible to achieve quite collaborative results. And, in fact, that was generally the case in the IMIX Programme as a whole, as discussed below.

In view of the programme's time frame, the output of the four main theme groups was sufficient, in terms of both quality and quantity. All research groups have an outstanding publication record for a project of this size. Most projects delivered an impressive amount of work for a small budget, leading to true 'value for money'. Each theme is discussed in turn.

Speech technology

The automatic speech recognition (ASR) research in NORISC was solid work. It provided the rejection of a plausible hypothesis about the value of using longer units in the recognition process. The fact that this strategy was evidently not as successful as was initially hoped is in itself a valid result that is on a par with current research in the area of ASR. NORISC was also supposed to produce ASR for the Demonstrator, but because of personnel changes the inclusion of the open source ASR system HTP was chosen instead. For the development of attractive Demonstrator systems, the Review Committee strongly encouraged the use of new open source software and the use of skills and products from technical support all over the world, from outside the funded projects and even outside of academia proper.

Question Answering

The QA projects all performed research of top quality, although the chosen approaches and topics were not particularly innovative or risky: really new ‘crazy’ ideas, often so typically Dutch, were missing. The research was concerned mainly with re-engineering existing approaches for Dutch and for the self-education of a new generation of researchers. This is of course very valuable in itself. Strong collaboration and close connection between all projects was observed, which the panel considered to be tangible evidence for the formation of a potentially very valuable network consisting of the next generation of talent in speech and language technology.

The QADR project[4] showed good research at current international levels, with a good balance of novel research and engineering or implementation. It revealed numerous publications and extensive collaboration with partners inside and outside IMIX. The research included syntax for passage-level IR, IE, and coreference for QA, learning extraction patterns using seeds, estimating semantic similarity of words for QA matching, learning synonyms, recognising definition sentences, and learning categories using appositives, and more. This impressive amount of work yielded solid and shared results in several of these areas and numerous publications.

The ROLAQUAD project[5] delivered two versions of a QA system for the Demonstrator, for question analysis, retrieval and answer ranking. The overall technical focus was on sophisticated processing methods and led to nice computational research and results. There were some collaboration discussions with other IMIX partners, but the main outreach was to international researchers. Serious efforts were performed to link with the QA company Textkernel.[6] Researchers of the Factmine project[7] published a remarkably large number of papers for such a small

[4] www.let.rug.nl/~gosse/Imix/

[5] http://ilk.uvt.nl/rolaquad/

[6] www.textkernel.nl

[7] http://ifarm.nl/erikt/factmine/

project. Their system supported QA and Information Extraction reasoning, question analysis, and more. Unfortunately, the final report provided relatively little detail about the system. It would have been useful to know what was learned after the project was refocused on learning only hypernyms from the open domain, instead of the medical domain. It would also have been helpful to know how the results lined up with EuroWordNet or English-based hypernym resources.

Dialogue management

The two projects on dialogue management followed complementary approaches, leading to two different Dialogue Managers, which made cooperation amongst the projects difficult. On the other hand, it also potentially enabled comparison of the two Managers, which unfortunately was not carried out. As regards the intended naturalness of the interaction, the Review Panel advised developing scenarios in which people really care for naturalness, or where interaction becomes severely disturbed by unnaturalness, and to distinguish these from situations in which naturalness does not matter much.

The major result of PARADIME[8] consists of a good contribution to theory of dialogue. The focus was on developing an architecture to support intentional (by reasoning) and social (scripted) dialogue management. In the developed model the 'evaluation agent' plays the hardest task.

VIDIAM[9] showed much work of various kinds, based on several collected and analysed dialogues and the use of a Dialogue Act recogniser, multimodal fusion, and a shared image database with IMOGEN.[10] The remaining key question here concerned the smartest way to perform discourse annotation. For image annotation it is advised to work together with research on image retrieval techniques.

Both projects produced several papers. The researchers in this field were strongly encouraged to come forward with their results to ensure the attention they deserve.

Multimodal output and speech synthesis

Especially for the multimodal I/O research in IMIX the results were very interesting, given the actual set-up of the projects.

IMOGEN dealt with various aspects of multimodal output generation and showed excellent work: serious thought and solid methodology were devoted to a challenging problem. The researchers performed extensive collaboration with other IMIX partners, as well as internationally. Moreover, they established a new separate workshop series, MOG (Multimodal Output Generation), which had significant international participation from well-respected researchers. One very interesting

[8] www.nwo.nl/nwohome.nsf/pages/NWOP_653H9J

[9] wwwhome.cs.utwente.nl/~schooten/vidiam/

[10] wwwhome.cs.utwente.nl/~theune/IMOGEN/

result of IMOGEN that will be useful well beyond the end of IMIX is the corpus of multimodal presentations collected in the doctoral dissertation of van Hooijdonk (2008).

This research in multimodal presentation using a talking head that expresses emotions in the facial modality was limited by the speech synthesis systems used, which were not able to express vocal emotions. Here a look at approaches as put forward in the framework of the EU-funded Humaine network might have helped. Future projects should preferably strive for coherence of emotion expression in all modalities, as from clinical studies it appears to be relevant in human natural perception and understanding. Overall, the multimodal output presentation work was on a par with the state of the art internationally.

2.2 Programmatic Evaluation

In general, the IMIX Programme was successful in meeting the general criteria described in the programme text and outlined in Chapter 1, i.e., the three strategic priorities:

(i) Strengthening the multidisciplinary collaboration and strengthening knowledge transfer between academia and industry;
(ii) Strengthening the position of the Dutch language in the Information Society;
(iii) Building a common demonstrator to improve the visibility of language and speech technology as enabler of advanced information services.

The IMIX Programme clearly gave an impulse to multidisciplinary collaboration and strengthened knowledge transfer through active collaboration between and interaction across projects, transfer of project members to other research projects within and outside the Netherlands, and transfer of some PhD students and post-docs from academia to industry.

The IMIX Programme ensured that state-of-the art language and speech technology was actively applied to the Dutch language. In all projects, the Dutch language was highlighted, notably in the QA system Joost (Bouma et al, 2005) developed in QADR, and in ROLAQUAD; in the other projects general-purpose developments were applied to Dutch.

From the written material (including the final report) and the presentations at the final workshop the panel concluded that the component teams were well chosen in the senses of being complementary (even in the case of the two projects on dialogue management, who took different approaches), of addressing important problems, and of each in their own ways contributing to a clear overall vision.

Many of the IMIX researchers have been playing prominent roles in language and speech technology for a number of years already. They were, for example, involved in many trans-national EU FP7 projects and also the presence of IMIX researchers in various international competitions is an indicator that they are present on the international ICT landscape.

2.3 Delivery and Outreach

The IMIX Programme was visible to the national and international research communities in speech and language technology, to relevant industry, and to the interested public. Overall, the programme resulted in a quite remarkable publication output given its small size. Compared to US or German standards, IMIX was really actually the size of a single medium/large project (between 400K and 500K euro per year). Seen in that light, the publication level was astounding. Every project had publications at a major conference. Some sub-areas of course have more centrality in terms of the main conferences than others; nonetheless, the projects have a group record in this regard. From this the Review Panel concluded that the scientific visibility of the mostly junior researchers employed within the IMIX Programme was clear and very good. However, a larger programme that allowed serious involvement of senior researchers would of course have been able to achieve an even greater impact on the international scene.

Both the connections to industry and the exchange of people and ideas were successful, especially considering the dire situation of Dutch and Flemish HLT industry at that moment. It is a pity that no patents or spin-off companies resulted from this work.

Concerning the visibility to the public, a single large press event might have been nice, but the absence of a central Demonstrator made this rather difficult. However, the IMIX film[11] about the construction and background of the Demonstrator is effective, and may well serve this goal in the near future.

3 Recommendations for the Future

As with almost all research endeavours, the programme's goals were more ambitious than the level of funding allowed. To fully achieve all the goals, the budget which would really have been needed is five to tenfold the actual amount. As it was, the teams covered a remarkable amount of territory in the general sphere of multimodal Question Answering and information delivery, including: speech and text; Question Answering and information extraction; component technologies, such as named entity recognition; dialogue management; and some multimodal presentation reasoning. Had there been more funding it would have been interesting to include geospatial information delivery (maps, satellite images, directions, etc.), work on multimodal question input (both 1D input such as menus and type-in, and 2D input such as pointing), user tailoring of presentation, etc. These are research and implementation directions that at the time of the publication of this volume are beginning to make inroads into the market, especially in the context of smartphones.

Both multidisciplinary and inter-university collaboration were achieved in the IMIX Programme. The programme clearly spurred the collaboration between the

[11] www.youtube.com/watch?v=swA8_Y56aak

important players in the Netherlands in the area of HLT. Many visits took place, both within and outside of the Netherlands. In addition, there seems to have been excellent discussion and considerable sharing across projects. As a result, IMIX was very successful in establishing a cadre of people, forming a strong network and using a shared language, that will prove of considerable value in the future.

The Demonstrator functioned as a vehicle to stimulate cooperation, not a true integration system. In that sense, it did succeed, but it imposed a certain overhead on participants, too. Had all projects just been asked to work on the same data, and encouraged to produce pair-wise project collaborations, perhaps the same level of cooperation might have been achieved more effectively. Something that could have been strengthened in this respect is joint corpus creation that supports multiple functionalities, modalities, or aspects on the same annotated resource(s).

IMIX delivered some new resources and improvements to existing ones (the Alpino parser[12], the TIMBL machine learning package[13], etc.), thereby realising some of the necessary parts of the infrastructure required for integrated multimedia processing. A new research programme, focusing on extending this kind of infrastructure to facilitate rapid prototyping of new experiments in multimedia and other domains, would be an extremely wise investment at the current time.

Three kinds of infrastructure are needed to support future multimedia information development:

1. Static resources: Corpora of multimedia questions and answers, information request dialogues, corpus creation and annotation tools, etc. A small quantity of such material is available internationally, but not for Dutch, and there is nothing that integrates language, speech, and images or maps.
2. Dynamic resources: A collection and standardisation of existing Dutch language, speech, and other medium processing technology, such as speech and language pre-processors, parsers, machine learning packages, generators, dialogue agents, Knowledge Representation systems, etc., with a common simple input and output formalism or APIs and clear instructions for downloading and usage.
3. Evaluations: Annotated and analysed datasets that can be used to perform standardised comparative evaluations of systems and parts of systems; the annotations must be made with regards to theory-neutral information delivery and cross-media integration functionality in mind. Given the complexity of evaluating multimedia information and of evaluating dialogue, this is a research programme in its own right. Considerable care should be taken to make sure that there is a balance between glass-box and black-box material and that the evaluations are not biased to any particular theory.

In order to validate, guide, and exploit a new infrastructure effort, a follow-on research programme should be planned as well. Such a programme should include both a common framework, e.g., based on pre-constructed corpora, tasks,

---

[12] `www.let.rug.nl/~vannoord/alp/Alpino/`

[13] `http://ilk.uvt.nl/timbl/`

and evaluation criteria and data, that allows different combinations of technologies and interaction paradigms to be rapidly configured, as well as a serious human factors evaluation component in which these configurations can be tested and new suggestions can be produced for further component or integration research. The Review Panel recommends that any future programme should be large enough to also include the participation of institutions in the area of Human Computer Interaction as well as industrial partners. Interactions and synergies between participants from areas that approach human communications from different angles should be actively encouraged.

For a small country, the Netherlands has always been one of the leaders in the general field of Human Language Technology, and IMIX is no exception. Multimodal information delivery is a dynamic and ever-growing research area. It continues to be greatly needed by society, as it provides the basis to interactively analyse, evaluate, and share information, and consequently may increase efficiency and productivity in industry and in scientific research and lead to new and innovative developments. The Netherlands can strengthen its leading position by establishing a new programme that builds upon IMIX.

References

Bouma G, Fahmi I, Mur J, van Noord G, van der Plas L, Tiedeman J (2005) Linguistic knowledge and question answering. Traitement Automatique des Langues 2(46):15–39

Hooijdonk, C M J van (2008) Explorations in Multimodal Information Presentation. PhD thesis, Tilburg University, The Netherlands